The ARRL's
Tech Q&A

Your Quick & Easy Path to a Technician Ham License

By:
Larry D. Wolfgang, WR1B and Joel P. Kleinman, N1BKE

Production Staff:
David Pingree, N1NAS, Senior Technical Illustrator:
 Technical Illustrations
Jayne Pratt-Lovelace, Proofreader
Paul Lappen, Production Assistant: Layout
Sue Fagan, Graphic Design Supervisor: Cover Design
Michelle Bloom, WB1ENT, Production Supervisor: Layout

Published By:
ARRL—The national association for Amateur Radio
Newington, CT 06111-1494

At Press Time

ARRL Files Partial Reconsideration Petition on Restructuring

The ARRL has formally asked the FCC to reconsider and modify two aspects of its December 30, 1999, Report and Order that restructured the Amateur Radio rules. The ARRL wants the FCC to continue to maintain records that indicate whether a Technician licensee has Morse code element credit. It also seeks permanent Morse element credit for any Amateur Radio applicant who has ever passed an FCC-recognized Morse exam of at least 5 WPM.

The ARRL filed a Petition for Partial Reconsideration in the WT Docket 98-143 proceeding on March 13, 2000.

The ARRL suggested that it would be less of an administrative burden for the FCC to maintain the Technician database as it has been doing. The database now identifies Technician and Tech Plus licensees by encoding the records with a "T" or a "P" respectively. The ARRL also said the inability to identify those Technicians who have HF privileges and those who do not could hamper voluntary enforcement efforts. It further suggested it would be wrong to put the burden of proof of having passed the Morse examination on licensees.

The ARRL cited the demands of fairness in asking the FCC to afford Morse element credit to all applicants who have ever passed an FCC-recognized 5 WPM code exam. The rules already grant Element 1 credit to those holding an expired or unexpired FCC-issued Novice license or an expired or unexpired Technician Class operator license document granted before February 14, 1991. It also grants Element 1 credit to applicants possessing an FCC-issued commercial radiotelegraph operator license or permit that's valid or expired less than 5 years. The ARRL has asked the FCC to "conform the rules" to give similar credit to those who once held General, Advanced or Amateur Extra class licenses.

Contents

iv   Foreword

vi   When to Expect New Books

vii   How to Use This Book

viii   ARRL Study Materials

1   Introduction

19   Technician Subelement 1
Commission's Rules

59   Technician Subelement 2
Operating Procedures

79   Technician Subelement 3
Radio-Wave Propagation

95   Technician Subelement 4
Amateur Radio Practice

115   Technician Subelement 5
Electrical Principles

129   Technician Subelement 6
Circuit Components

139   Technician Subelement 7
Practical Circuits

151   Technician Subelement 8
Signals and Emissions

159   Technician Subelement 9
Antennas and Feed Lines

169   Technician Subelement 10
RF Safety

184   About the ARRL

Congratulations! You have taken the first step into the most exciting pastime in the world! With your Amateur Radio license in hand, you can communicate with other hams—in your town or city, in your part of your state or around the world!

No matter what other interests you may have, you will find ways to tie them to Amateur Radio.

♦ *Computers?* Hams have been using computers to enhance their enjoyment of their hobby since the 1970s! Computer-based *data modes* are among the fastest growing means of Amateur Radio communication. Hams also use computers to design and analyze their antenna systems, and to build and test radio equipment and accessories.

♦ *Video?* Using computer software and video cameras, hams exchange moving and still images via Amateur Television or Slow Scan TV.

♦ *Electronics?* There's no better way to learn about electronics than to prepare for a license exam or to build a useful piece of equipment for your Amateur Radio station. Some hams communicate around the world with only *5 watts of power.*

♦ *Public service?* Although ham radio operators do not (and in fact can not) accept payment for the communications services they provide, most hams are involved in community service in some way. Many volunteers enjoy supplying communications to walkathons and similar events. Others take part in government-sponsored emergency management exercises to help prepare for the real thing. It's been proven time and again that when the power goes out and phone lines are dead, there's no better means of communication than ham radio.

♦ *Hiking, biking and other outdoor activities?* Many hams love nothing more than packing a wire antenna and a small, lightweight radio and heading for the great outdoors. You can almost always find someone to talk to from anyplace, but if you're up high, so much the better—you'll extend your range significantly.

ARRL, the national association for Amateur Radio, has been producing materials for hams and prospective hams for more than 70 years. This book is among the many books, tapes, videos, software and CDs that ARRL offers to those preparing for an Amateur Radio exam.

Inside the pages of this book, you'll find everything you need to prepare for the Technician license exam. There are nearly 400 questions in the Technician question pool. This book provides the exact questions, answers and distracters (wrong answers) that will appear on your exam. There's a brief and accurate explanation for each question.

ARRL is pleased that you've chosen an ARRL publication to prepare for your first Amateur Radio license exam. We invite all those interested in Amateur Radio communication to join ARRL. An Amateur Radio license is not necessary. We have included a Membership Application elsewhere in this book for your convenience. Nearly 170,000 ARRL members enjoy a wide range of membership benefits, including a monthly magazine, *QST*, which offers articles and columns for beginners as well as more experienced hams.

There is also a Feedback Form at the back of the book. We'd like to hear from you—your comments and suggestions are important to us. Thanks, and good luck!

David Sumner, K1ZZ
Executive Vice President, ARRL
Newington, Connecticut
April 2000

When to Expect New Books

A Question Pool Committee (QPC) consisting of representatives from the various Volunteer Examiner Coordinators (VECs) writes amateur-licensing questions. The QPC establishes a schedule for revising and implementing new Question Pools.

The current QPC revision schedule is as follows:

Exam	*Question Pool Valid Through*
Technician (Element 2)	June 30, 2003
General (Element 3)	June 30, 2004
Extra Class (Element 4)	June 30, 2002

As new Question Pools are released, ARRL will produce new study materials before the effective dates of the new Pools. Until then, the current Question Pools will remain in use, and current ARRL study materials, including this book, will help you prepare for your exams.

As the new Question Pool schedules are confirmed, the information will be published in *QST* and on *ARRLWeb* at **http://www.arrl.org/**.

New Ham Desk
ARRL Headquarters
225 Main Street
Newington, CT 06111-1494
(860) 594-0200

Prospective new amateurs call:
800-32-NEW-HAM (800-326-3942)

You can also contact us via e-mail: **newham@arrl.org**
or check out our World Wide Web site:
http://www.arrl.org/

How to Use
This Book

To earn a Technician Amateur Radio license, you must pass the Technician written exam, FCC Element 2. You will have to know some basic electronics theory and Amateur Radio operating practices and procedures. In addition, you will need to learn about some of the rules and regulations governing the Amateur Service, as contained in Part 97 of Title 47 of the Code of Federal Regulations — the Federal Communications Commission (FCC) Rules.

The Element 2 exam consists of 35 questions about Amateur Radio rules, theory and practice. A passing score is 74%, so you must answer 26 of the 35 questions correctly to pass. (Another way to look at this is that you can get as many as 9 questions wrong, and still pass the test.)

The questions and multiple-choice answers in this book are printed exactly as they were written by the Volunteer Examiner Coordinator's Question Pool Committee, and exactly as they will appear on your exam. (Be careful, though. The letter positions of the answers may be scrambled, so you can't simply memorize an answer letter for each question.) In this book, the letter of the correct answer is printed in **boldface type** just before the explanation. If you want to study without knowing the correct answer right away, simply cover the answer letters with your hand or a slip of paper.

As you read the explanations for many of the questions you will find words printed in **boldface type**. These words are important terms, and will help you identify the correct answer to the question.

ARRL Study Materials

To earn a Technician license with Morse code privileges, you'll have to be able to send and receive the international Morse code at a rate of 5 wpm. This license gives you operating privileges on the high-frequency (HF) bands, and offers the excitement of worldwide communications.

ARRL offers *Your Introduction to Morse Code*, a set of cassette tapes or audio CDs that teach you the code. The two cassettes or audio CDs introduce each of the required characters and provide practice on each character as it is introduced. Then the character is used in words and text before proceeding to the next character. After all characters have been introduced, there is plenty of practice at 5 wpm to help you prepare for the Element 1 (5 wpm) exam.

ARRL also offers code-practice sets to help you increase your speed: *Increasing Your Code Speed: 5 to 10 WPM, 10 to 15 WPM* and *15 to 22 WPM*. Each set includes two cassette tapes or two audio CDs with a variety of practice at gradually increasing speeds over the range for that set. The Morse code on these practice sets is sent using 18-wpm characters, with extra space between characters to slow the overall code speed, for code up to 18 wpm. This same technique is used on ARRL/VEC Morse code exams.

For those who prefer a computer program to learn and practice Morse code, ARRL offers the excellent GGTE *Morse Tutor Gold* program for IBM PC and compatible computers.

Even with the tapes, CDs or computer program, you'll want to tune in the code-practice sessions transmitted by W1AW, the ARRL Headquarters station. For more information about W1AW or how to order any ARRL publication or set of code tapes, write to ARRL Headquarters, 225 Main St, Newington, CT 06111-1494, tel (toll free) 1-888-277-5289. You can also order at *ARRLWeb*, **http://www.arrl.org**.

JOIN ARRL TODAY AND RECEIVE A *FREE* BOOK!

I want to join ARRL. Send me the *FREE* book I have selected (choose one):

☐ ***Repeater Directory***—gives you listings of more than 20,000 FM voice repeaters throughout the US. ($8 value)

☐ ***Your VHF Companion***—lets you explore the fascinating activities on the VHF bands: FM, repeaters, packet, CW, SSB, satellites, amateur television, and more. ($10 value)

☐ New member   ☐ Previous member   ☐ Renewal

_____

Call Sign (if any)   Class of License   Date of Birth

_____

Name

_____

Address

_____

City, State, ZIP

Dues are $34 in US / $47 in Canada (US funds) / $54 elsewhere (US funds). You do not need an Amateur Radio license to join. Individuals age 65 or over, residing in the US, upon submitting one-time proof of age, may request the dues rate of $28. Immediate relatives of a member who receives *QST*, and reside at the same address may request family membership at $5 per year. Blind individuals may join without *QST* for $5 per year. If you are 21 or younger and a licensed amateur, a special rate may apply. Write to ARRL for details.

Sorry! Free book offer does not apply to individuals joining as family or blind members or submitting their application via clubs.

DUES ARE SUBJECT TO CHANGE WITHOUT NOTICE.

Payment Enclosed ☐

Charge to MC, VISA, AMEX, Discover No. _____

Expiration Date _____

Cardholder Name _____

Cardholder Signature _____

If you do not wish your name and address made available for non-ARRL related mailings, please check this box ☐

ARRL
225 MAIN STREET NEWINGTON, CONNECTICUT 06111 USA
New Hams call (800) 326-3942
Call toll free to join: (888) 277-5289
Join on the Web: http://www.arrl.org/join.html   TQA 00

Introduction

The Technician License

Earning a Technician Amateur Radio license is a good way to begin enjoying ham radio. There is no Morse code exam for this license, and the Element 2 written exam is straightforward, with no difficult math or electronics background required. You are sure to find the operating privileges available to a Technician licensee to be worth the time spent learning about Amateur Radio. After passing the exam, you will be able to operate on *every frequency* above 50 megahertz that is assigned to the Amateur Radio Service. With full operating privileges on those bands, you'll be ready to experience the excitement of Amateur Radio!

Perhaps you are mainly interested in local communications using FM repeaters. Maybe you want to use your computer to explore the many digital modes of communication. If your eyes turn to the stars on a clear night, you might enjoy tracking the amateur satellites and using them to relay your signals to other amateurs around the world!

Once you make the commitment to study and learn what it takes to pass the exam, you *will* accomplish your goal. Many people pass the exam on their first try, so if you study the material and are prepared, chances are good that you will soon have your license. It may take you more than one attempt to pass the Technician license exam, but that's okay. There is no limit to how many times you can take it. Many Volunteer Examiner Teams have several exam versions available, so you may even be able to try the exam again at the same exam session. Time and available exam versions may limit the number of times you can try the exam at a single exam session. If you don't pass after a couple of tries you will certainly benefit from more study of the question pools before you try again.

An Overview of Amateur Radio

Earning an Amateur Radio license, at whatever level, is a special achievement. The 600,000 or so people in the US who call themselves Amateur Radio operators, or hams, are part of a global fraternity. Radio amateurs provide a voluntary, noncommercial, communication service. This is especially true during natural disasters or other emergencies. Hams have made many important contributions to the field of electronics and communications, and this tradition continues today. Amateur Radio experimentation is yet another reason many people become part of this self-disciplined group of trained operators, technicians and electronics experts — an asset to any country. Hams pursue their hobby purely for personal enrichment in technical and operating skills, without any type of payment except the personal satisfaction they feel from a job well done!

Radio signals do not know territorial boundaries, so hams have a unique ability to enhance international goodwill. Hams become ambassadors of their country every time they put their stations on the air.

Amateur Radio has been around since before World War I, and hams have always been at the forefront of technology. Today, hams relay signals through their own satellites, bounce signals off the moon, relay messages automatically through computerized radio networks and use any number of other "exotic" communications techniques. Amateurs talk from hand-held transceivers through mountaintop repeater stations that can relay their signals to other hams' cars or homes. Hams send their own pictures by television, talk with other hams around the world by voice or, keeping alive a distinctive traditional skill, tap out messages in Morse code. When emergencies arise, radio amateurs are on the spot to relay information to and from disaster-stricken areas that have lost normal lines of communication.

The US government, through the Federal Communications Commission (FCC), grants all US Amateur Radio licenses. This licensing procedure ensures operating skill and electronics know-how. Without this skill, radio operators, because of improperly adjusted equipment or neglected regulations, might unknowingly cause interference to other services using the radio spectrum.

Who Can Be a Ham?

The FCC doesn't care how old you are or whether you're a US citizen. If you pass the examination, the Commission will issue you an amateur license. Any person (except the agent of a foreign government) may take the exam and, if successful, receive an amateur license. It's important to understand that if a citizen of a foreign country receives an amateur license in this manner, he or she is a US Amateur Radio operator. (This should not be confused with a reciprocal permit for alien amateur licensee,

Figure 1—With a Technician Amateur Radio license, you'll be able to use hand-held radios to operate over FM repeaters on the VHF and UHF bands.

which allows visitors from certain countries who hold valid amateur licenses in their homelands to operate their own stations in the US without having to take an FCC exam.)

License Structure

Anyone earning a new Amateur Radio license can earn one of three license classes — Technician, General and Amateur Extra. These vary in degree of knowledge required and frequency privileges granted. Higher class licenses have more comprehensive examinations. In return for passing a more difficult exam you earn more frequency privileges (frequency space and modes of operation). The vast majority of beginners start with the most basic license, the Technician, although it's possible to start with any class of license.

Technician licensees who learn the international Morse code and pass an exam to demonstrate their knowledge of code at 5 wpm gain some frequency privileges on four of the amateur high-frequency (HF) bands. This license was previously called the Technician Plus license, and many amateurs will refer to it by that name. **Table 1** lists the amateur license classes you can earn, along with a brief description of their exam requirements and operating privileges.

Although there are also other amateur license classes, the FCC is no longer issuing new licenses for these classes. The Novice license was long considered the beginner's license. Exams for this license were discontinued as of April 15, 2000. The FCC also stopped issuing new Advanced class licenses on that date. They will continue to renew previously issued licenses, however, so you will probably meet some Novice and Advanced class licensees on the air.

The written Technician exam, called Element 2, covers some basic radio fundamentals and knowledge of some of the rules and regulations in Part 97 of the FCC Rules. With a little study you'll soon be ready to pass the Technician exam.

Each step up the Amateur Radio license ladder requires the applicant to pass the lower exams. So if you want to start out as a General

---

Table 1

Amateur Operator Licenses†

Class	Code Test	Written Examination	Privileges
Technician		Basic theory and regulations. (Element 2)*	All amateur privileges above 50.0 MHz.
Technician with Morse code credit	5 wpm (Element 1)	Basic theory and regulations. (Element 2)*	All "Novice" HF privileges in addition to all Technician privileges.
General	5 wpm (Element 1)	Basic theory and regulations; General theory and regulations. (Elements 2 and 3)	All amateur privileges except those reserved for Advanced and Amateur Extra class.
Amateur Extra	5 wpm (Element 1)	All lower exam elements, plus Extra-class theory (Elements 2, 3 and 4)	All amateur privileges.

†A licensed radio amateur will be required to pass only those elements that are not included in the examination for the amateur license currently held.

*If you have a Technician-class license issued before March 21, 1987, you also have credit for Elements 1 and 3. You must be able to prove your Technician license was issued before March 21, 1987 to claim this credit.

class or even an Amateur Extra class licensee, you must also pass the Technician written exam.

Anyone (except an agent or representative of a foreign government) is eligible to qualify for an Amateur Radio operator license. There is no age requirement. A Technician license gives you the freedom to develop operating and technical skills through on-the-air experience. These skills will help you upgrade to a higher class of license, with additional privileges.

As a Technician, you can use a wide range of frequency bands — *all amateur bands above 30 MHz*, in fact. You'll be able to use point-to-point communications on VHF FM, and repeaters, packet radio and orbiting satellites to relay your signals over a wider area. You can provide public service through emergency communications and message handling.

By passing the 5-wpm Morse code test you will have a *Technician license with Morse code credit*. With this license you will gain frequency privileges on four of the HF bands. See **Figure 2**. On those bands you will experience the thrill of *working* (contacting) other Amateur Radio operators in just about any country in the world. There's nothing quite like making friends with other amateurs around the world.

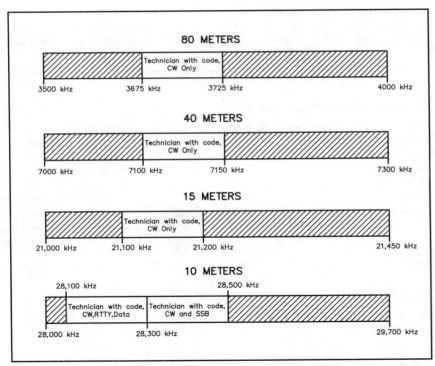

Figure 2.

Table 2
Amateur Operating Privileges

US Amateur Bands
April 15, 2000
Novice, Advanced and Technician Plus Allocations

New Novice, Advanced and Technician Plus licenses will not be issued after April 15, 2000. However, the FCC has allowed the frequency allocations for these license classes to remain in effect. They will continue to renew existing licenses for those classes.

160 METERS

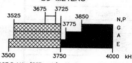

E,A,G

1800   1900   2000   kHz

Amateur stations operating at 1900—2000 kHz must not cause harmful interference to the radiolocation service and are afforded no protection from radiolocation operations.

80 METERS

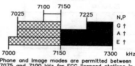

3525   3675   3725   3850   N,P
                3775   G
                                 A
                                 E

3500   3750   4000   kHz

5167.5 kHz (SSB only): Alaska emergency use only.

40 METERS

7025   7100   7150   7225   N,P †
                                 G †
                                 A †
                                 E †

7000   7150   7300   kHz

† Phone and image modes are permitted between 7075 and 7100 kHz for FCC licensed stations in ITU Regions 1 and 3 and by FCC licensed stations in ITU Region 2 West of 130 degrees West longitude or South of 20 degrees North latitude. See Sections 97.305(c) and 97.307(f)(11). Novice and Technician Plus licensees outside ITU Region 2 may use CW only between 7050 and 7075 kHz. See Section 97.301(e). These exemptions do not apply to stations in the continental US.

30 METERS

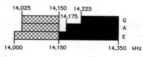

E,A,G

10,100   10,150   kHz

Maximum power on 30 meters is 200 watts PEP output. Amateurs must avoid interference to the fixed service outside the US.

20 METERS

14,025   14,150   14,225   G
                    14,175   A
                               E

14,000   14,150   14,350   kHz

17 METERS

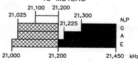

E,A,G

18,068   18,110   18,168   kHz

15 METERS

21,025   21,100   21,200   21,300   N,P
                              21,225   G
                                         A
                                         E

21,000   21,200   21,450   kHz

12 METERS

E,A,G

24,890   24,930   24,990   kHz

10 METERS

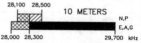

28,100   28,500   N,P
                          E,A,G

28,000   28,300   29,700   kHz

Novices and Technician Plus licensees are limited to 200 watts PEP output on 10 meters.

6 METERS

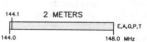

50.1   E,A,G,P,T
50.0   54.0   MHz

2 METERS

144.1   E,A,G,P,T
144.0   148.0   MHz

1.25 METERS

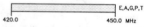

E,A,G,P,T,N

222.0   225.0   MHz

Novices are limited to 25 watts PEP output from 222 to 225 MHz.

70 CENTIMETERS **

E,A,G,P,T

420.0   450.0   MHz

33 CENTIMETERS **

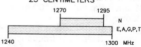

E,A,G,P,T

902.0   928.0   MHz

23 CENTIMETERS **

1270   1295   N
                      E,A,G,P,T

1240   1300   MHz

Novices are limited to 5 watts PEP output from 1270 to 1295 MHz.

US AMATEUR POWER LIMITS

At all times, transmitter power should be kept down to that necessary to carry out the desired communications. Power is rated in watts PEP output. Unless otherwise stated, the maximum power output is 1500 W. Power for all license classes is limited to 200 W in the 10,100—10,150 kHz band and in all Novice subbands below 28,100 kHz. Novices and Technicians with Morse code credit are restricted to 200 W in the 28,100—28,500 kHz subbands. In addition, Novices are restricted to 25 W in the 222—225 MHz band and 5 W in the 1270—1295 MHz subband.

Operators with Technician class licenses and above may operate on all bands above 50 MHz. For more detailed information see The FCC Rule Book.

— KEY —

▨ = CW, RTTY and data

☐ = CW, RTTY, data, MCW, test, phone and image

■ = CW, phone and image

▨ = CW and SSB phone

▨ = CW, RTTY, data, phone, and image

☐ = CW only

E = EXTRA CLASS
A = ADVANCED
G = GENERAL
P = TECHNICIAN PLUS
T = TECHNICIAN
N = NOVICE

* Effective April 15, 2000, Technicians passing the Morse code exam will gain HF Novice privileges, although they still hold a Technician license.

** Geographical and power restrictions apply to these bands. See The FCC Rule Book for more information about your area.

Above 23 Centimeters:

All licensees except Novices are authorized all modes on the following frequencies:
2300—2310 MHz
2390—2450 MHz
3300—3500 MHz
5650—5925 MHz
10.0—10.5 GHz
24.0—24.25 GHz
47.0—47.2 GHz
75.5—81.0 GHz
119.98—120.02 GHz
142—149 GHz
241—250 GHz
All above 300 GHz

For band plans and sharing arrangements, see The ARRL Operating Manual or The FCC Rule Book.

Table 3

W1AW SCHEDULE

Pacific	Mtn	Cent	East	Mon	Tue	Wed	Thu	Fri
6 AM	7 AM	8 AM	9 AM		Fast Code	Slow Code	Fast Code	Slow Code
7 AM-1 PM	8 AM-2 PM	9 AM-3 PM	10 AM-4 PM	Visiting Operator Time (12 PM - 1 PM closed for lunch)				
1 PM	2 PM	3 PM	4 PM	Fast Code	Slow Code	Fast Code	Slow Code	Fast Code
2 PM	3 PM	4 PM	5 PM	Code Bulletin				
3 PM	4 PM	5 PM	6 PM	Teleprinter Bulletin				
4 PM	5 PM	6 PM	7 PM	Slow Code	Fast Code	Slow Code	Fast Code	Slow Code
5 PM	6 PM	7 PM	8 PM	Code Bulletin				
6 PM	7 PM	8 PM	9 PM	Teleprinter Bulletin				
6⁴⁵ PM	7⁴⁵ PM	8⁴⁵ PM	9⁴⁵ PM	Voice Bulletin				
7 PM	8 PM	9 PM	10 PM	Fast Code	Slow Code	Fast Code	Slow Code	Fast Code
8 PM	9 PM	10 PM	11 PM	Code Bulletin				

W1AW's schedule is at the same local time throughout the year. The schedule according to your local time will change if your local time does not have seasonal adjustments that are made at the same time as North American time changes between standard time and daylight time. From the first Sunday in April to the last Sunday in October, UTC = Eastern Time + 4 hours. For the rest of the year, UTC = Eastern Time + 5 hours.

• **Morse code transmissions:**
Frequencies are 1.818, 3.5815, 7.0475, 14.0475, 18.0975, 21.0675, 28.0675 and 147.555 MHz.
Slow Code = practice sent at 5, 7$\frac{1}{2}$, 10, 13 and 15 wpm.
Fast Code = practice sent at 35, 30, 25, 20, 15, 13 and 10 wpm.
Code practice text is from the pages of *QST*. The source is given at the beginning of each practice session and alternate speeds within each session. For example, "Text is from July 1992 *QST*, pages 9 and 81," indicates that the plain text is from the article on page 9 and mixed number/letter groups are from page 81.
Code bulletins are sent at 18 wpm.
W1AW qualifying runs are sent on the same frequencies as the Morse code transmissions. West Coast qualifying runs are transmitted on approximately 3.590 MHz by K6YR. At the beginning of each code practice session, the schedule for the next qualifying run is presented. Underline one minute of the highest speed you copied, certify that your copy was made without aid, and send it to ARRL for grading. Please include your name, call sign (if any) and complete mailing address. Send a 9¥12-inch SASE for a certificate, or a business-size SASE for an endorsement.

• **Teleprinter transmissions:**
Frequencies are 3.625, 7.095, 14.095, 18.1025, 21.095, 28.095 and 147.555 MHz.
Bulletins are sent at 45.45-baud Baudot and 100-baud AMTOR, FEC Mode B. 110-baud ASCII will be sent only as time allows.
On Tuesdays and Fridays at 6:30 PM Eastern Time, Keplerian elements for many amateur satellites are sent on the regular teleprinter frequencies.

• **Voice transmissions:**
Frequencies are 1.855, 3.99, 7.29, 14.29, 18.16, 21.39, 28.59 and 147.555 MHz.

• **Miscellanea:**
On Fridays, UTC, a DX bulletin replaces the regular bulletins.
W1AW is open to visitors from 10 AM until noon and from 1 PM until 3:45 PM on Monday through Friday. FCC licensed amateurs may operate the station during that time. Be sure to bring your current FCC amateur license or a photocopy.
In a communication emergency, monitor W1AW for special bulletins as follows: voice on the hour, teleprinter at 15 minutes past the hour, and CW on the half hour.
Headquarters and W1AW are closed on New Year's Day, President's Day, Good Friday, Memorial Day, Independence Day, Labor Day, Thanksgiving and the following Friday, and Christmas

Learning Morse Code

Even if you don't plan to use Morse code now, there may come a time when you decide you would like to upgrade your license and earn those HF privileges. Learning Morse code is a matter of practice. Instructions on learning the code, how to handle a telegraph key, and so on, can be found in *Now You're Talking!*, published by the ARRL. In addition, *Your Introduction to Morse Code*, ARRL's package to teach Morse code, is available with two cassette tapes or two audio CDs. *Your Introduction to Morse Code* was designed for beginners, and will help you learn the code. *Increasing Your Code Speed* is a series of additional cassettes or audio CDs for code practice at speeds of 5 to 10, 10 to 15 and 15 to 22 wpm. You can purchase any of these products from your local Amateur Radio equipment dealer or directly from the ARRL, 225 Main St, Newington, CT 06111. To place an order, call, toll-free, **888-277-5289**. You can also send e-mail to: **pubsales@arrl.org** or check out our World Wide Web site: **http://www.arrl.org/** Prospective new amateurs can call: **800-32-NEW HAM** (**800-326-3942**) for additional information.

Besides listening to code tapes or CDs, some on-the-air operating experience will be a great help in building your code speed. When you are in the middle of a contact via Amateur Radio, and have to copy the code the other station is sending to continue the conversation, your copying ability will improve quickly! Although you did not have to pass a Morse code test to earn your Technician license, there are no regulations prohibiting you from using code on the air. Many amateurs operate code on the VHF and UHF bands.

ARRL's Maxim Memorial Station, W1AW, transmits code practice and information bulletins of interest to all amateurs. These code-practice sessions and Morse code bulletins provide an excellent opportunity for code practice. **Table 3** is a W1AW operating schedule. When we change from standard time to daylight saving time, the same local times are used.

Station Call Signs

Many years ago, by international agreement, the nations of the world decided to allocate certain call-sign prefixes to each country. This means that if you hear a radio station call sign beginning with W or K, for example, you know the station is licensed by the United States. A call sign beginning with the letter G is licensed by Great Britain, and a call sign beginning with VE is from Canada. *The ARRL DXCC List* is an operating aid no ham who is active on the HF bands should be without. That booklet, available from the ARRL, includes the common call-sign prefixes used by amateurs in virtually every location in the world. It also includes a check-off list to help you keep track of the countries you contact as you work toward collecting QSL cards from 100 or more countries to earn the prestigious DX Century Club award. (DX is ham lingo for distance, generally taken on the HF bands to mean any country outside the one from which you are operating.)

The International Telecommunication Union (ITU) radio regulations outline the basic principles used in forming amateur call signs. According to these regulations, an amateur call sign must be made up of one or two characters (the first one may be a numeral) as a prefix, followed by a numeral, and then a suffix of not more than three letters. The prefixes W, K, N and A are used in the United States. When the letter A is used in a US amateur call sign, it will always be with a two-letter prefix, AA to AL. The continental US is divided into 10 Amateur Radio call districts (sometimes called areas), numbered 0 through 9. **Figure 3** is a map showing the US call districts.

For information on the FCC's call-sign assignment system, and a table listing the blocks of call signs for each license class, see the ARRL publication, *The FCC Rule Book*. You may keep the same call sign when you change license class, if you wish. You must indicate that you want to receive a new call sign when you fill out an FCC Form 605 to apply for the exam or change your address.

The FCC also has a vanity call sign system. Under this system the FCC will issue a call sign selected from a list of preferred available call signs. While there is no fee for an Amateur Radio license, there is a fee for the selection of a vanity call sign. The current fee is $14 for a 10-year Amateur Radio license, paid upon application for a vanity call sign and at license renewal after that. (That fee may change as costs of administering the program change.) The latest details about the vanity call sign system are available from ARRL Regulatory Information, 225 Main Street, Newington, CT 06111-1494 and on *ARRLWeb* at **http://www.arrl.org/**

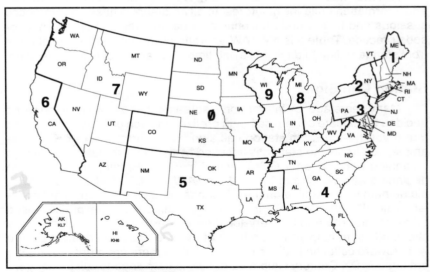

Figure 3—There are 10 US call areas. Hawaii is part of the sixth call area, and Alaska is part of the seventh.

EARNING A LICENSE

Forms and Procedures

To renew or modify a license, you can file a copy of FCC Form 605. In addition, hams who have held a valid license that has expired within the past two years may apply for reinstatement with an FCC Form 605.

Licenses are normally good for ten years. Your application for a license renewal must be submitted to the FCC no more than 90 days before the license expires. (We recommend you submit the application for renewal between 90 and 60 days before your license expires.) If the FCC receives your renewal application before the license expires, you may continue to operate until your new license arrives, even if it is past the expiration date. If you forget to apply before your license expires, you may still be able to renew your license without taking another exam. There is a two-year grace period, during which you may apply for renewal of your expired license. Use an FCC Form 605 to apply for reinstatement (and your old call sign). If you apply for reinstatement of your expired license under this two-year grace period, you may not operate your station until your new license is issued. If you move or change addresses you should use an FCC Form 605 to notify the FCC of the change. If your license is lost or destroyed, however, just write a letter to the FCC explaining why you are requesting a new copy of your license.

You can ask one of the Volunteer Examiner Coordinators' offices to file your renewal application electronically if you don't want to mail the form to the FCC. You must still mail the form to the VEC, however. The ARRL/VEC Office will electronically file application forms for any ARRL member free of charge.

You can also file your license renewal or address modification using the Universal Licensing System (ULS) on the World Wide Web. Go to **http://www.fcc.gov/wtb/uls** and click on the "TIN/Call Sign Registration" button. Follow the directions to register with the Universal Licensing System. Next click on the "Connecting to ULS" button and follow the directions given there to connect to the FCC's ULS database.

The FCC has a set of detailed instructions for the Form 605, which are included with the form. To obtain a new Form 605, call the FCC Forms Distribution Center at 800-418-3676. You can also write to: Federal Communications Commission, Forms Distribution Center, 2803 52nd Avenue, Hyattsville, MD 20781 (specify "Form 605" on the envelope). The Form 605 also is available from the FCC's fax on demand service. Call 202-418-0177 and ask for form number 000605. Form 605 also is available via the Internet. The World Wide Web location is: **http://www.fcc.gov/formpage.html** or you can receive the form via ftp to: **ftp.fcc.gov/pub/Forms/Form605**.

The ARRL/VEC has created a package that includes the portions of Form 605 that are needed for amateur applications, as well as a condensed set of instructions for completing the form. Write to: ARRL/VEC, Form 605, 225 Main Street, Newington, CT 06111-1494. (Please

NCVEC QUICK-FORM 605 APPLICATION FOR
AMATEUR OPERATOR/PRIMARY STATION LICENSE

SECTION 1 - TO BE COMPLETED BY APPLICANT

PRINT LAST NAME	SUFFIX	FIRST NAME	INITIAL	STATION CALL SIGN (IF ANY)
Dalton		Helen	W	

MAILING ADDRESS (Number and Street or P.O. Box)	SOCIAL SECURITY NUMBER / TIN (OR FCC LICENSEE ID #)
225 Main Street	069-30-4278

CITY	STATE CODE	ZIP CODE (5 or 9 Numbers)	E-MAIL ADDRESS (OPTIONAL)
Newington	CT	06111	

DAYTIME TELEPHONE NUMBER (Include Area Code) OPTIONAL	FAX NUMBER (Include Area Code) OPTIONAL	ENTITY NAME (IF CLUB, MILITARY RECREATION, RACES)

Type of Applicant: ☒ Individual ☐ Amateur Club ☐ Military Recreation ☐ RACES (Modify Only)

TRUSTEE OR CUSTODIAN CALL SIGN

I HEREBY APPLY FOR (Make an X in the appropriate box(es))

SIGNATURE OF RESPONSIBLE CLUB OFFICIAL

☒ **EXAMINATION** for a **new** license grant

☐ **EXAMINATION** for **upgrade** of my license class

☐ **CHANGE** my **name** on my license to my new name

Former Name: _____
(Last name) (Suffix) (First name) (MI)

☐ **CHANGE** my mailing address to **above** address

☐ **CHANGE** my station **call sign** systematically

Applicant's Initials: _____

☐ **RENEWAL** of my license grant.

Do you have another license application on file with the FCC which has not been acted upon?	PURPOSE OF OTHER APPLICATION	PENDING FILE NUMBER (FOR VEC USE ONLY)

I certify that:
* I waive any claim to the use of any particular frequency regardless of prior use by license or otherwise;
* All statements and attachments are true, complete and correct to the best of my knowledge and belief and are made in good faith;
* I am not a representative of a foreign government;
* I am not subject to a denial of Federal benefits pursuant to Section 5301of the Anti-Drug Abuse Act of 1988, 21 U.S.C. § 862;
* The construction of my station will NOT be an action which is likely to have a significant environmental effect (See 47 CFR Sections 1.301-1.319 and Section 97.13(a));
* I have read and WILL COMPLY with Section 97.13(c) of the Commission's Rules regarding RADIOFREQUENCY (RF) RADIATION SAFETY and the amateur service section of OST/OET Bulletin Number 65.

Signature of applicant (Do not print, type, or stamp. Must match applicant's name above.)

X Helen W. Dalton   Date Signed: May 20, 2000

SECTION 2 - TO BE COMPLETED BY ALL ADMINISTERING VEs

Applicant is qualified for operator license class:

☐ **NO NEW LICENSE OR UPGRADE WAS EARNED**

☒ **TECHNICIAN**   Element 2

☐ **GENERAL**   Elements 1, 2 and 3

☐ **AMATEUR EXTRA**   Elements 1, 2, 3 and 4

DATE OF EXAMINATION SESSION
5/20/00

EXAMINATION SESSION LOCATION
Newington, CT

VEC ORGANIZATION
ARRL

VEC RECEIPT DATE
COMPLETED MAY 2 2 2000

I CERTIFY THAT I HAVE COMPLIED WITH THE ADMINISTERING VE REQUIRMENTS IN PART 97 OF THE COMMISSION'S RULES AND WITH THE INSTRUCTIONS PROVIDED BY THE COORDINATING VEC AND THE FCC.

1st VEs NAME (Print First, MI, Last, Suffix)	VEs STATION CALL SIGN	VEs SIGNATURE (Must match name)	DATE SIGNED
MARTIN G. COOK	N1FOC	Martin G. Cook	5/20/2000
2nd VEs NAME (Print First, MI, Last, Suffix)	VEs STATION CALL SIGN	VEs SIGNATURE (Must match name)	DATE SIGNED
DANIEL P MILLER	K3UFG	Daniel P. Miller	5/20/2000
3rd VEs NAME (Print First, MI, Last, Suffix)	VEs STATION CALL SIGN	VEs SIGNATURE (Must match name)	DATE SIGNED
Larry D. Wolfgang	WR1B	Larry D. Wolfgang	5/20/00

NCVEC FORM 605 - APRIL 2000
FOR VE/VEC USE ONLY - Page 1

Figure 4—A completed NCVEC Quick Form 605 as it would be completed for a new Technician license.

ARE WRITTEN TESTS AN FCC-LICENSE REQUIREMENT? ARE THERE EXEMPTIONS?

Beginning April 15, 2000, you may be examined on only three classes of operator licenses, each authorizing varying levels of privileges. The class for which each examinee is qualified is determined by the degree of skill and knowledge in operating a station that the examinee demonstrates to volunteer examiners (VEs) in his or her community. The demonstration of this know-ledge is required in order to obtain an Amateur Operator/Primary Station License. There is no exemption from the written exam requirements for persons with difficulty in reading, writing, or because of a handicap or disability. There are exam accommo-dations that can be afforded examinees (see ACCOMMODATING A HANDICAPPED PERSON below). Most new amateur operators start at the Technician class and then advance one class at a time. The VEs give examination credit for the license class currently (and in some cases, previously) held so that examinations required for that license need not be repeated. The written examinations are constructed from question pools that have been made public (see: <http://www.arrl.org/arrlvec/pools.html>.) Helpful study guides and training courses are also widely avail-able. To locate examination opportunities in your area, contact your local club, VE group, one of the 14 VECs or see the online listings at: <http://www.w5yi.org/vol-exam.htm> or <http//www.arrl.org/arrlvec/examsearch.phtml>.

IS KNOWLEDGE OF MORSE CODE AN FCC-LICENSE REQUIREMENT? ARE THERE EXEMPTIONS?

Some persons have difficulty in taking Morse code tests because of a handicap or disability. There is available to all otherwise qualified persons, handicapped or not, the Technician Class operator license that does not require passing a Morse code examination. Because of international regulations, how-ever, any US FCC licensee seeking access to the HF bands (frequencies below 30 MHz) must have demonstrated proficiency in Morse code. If a US FCC licensee wishes to gain access to the HF bands, there is no exemption available from this Morse code proficiency requirement. If licensed as a Tech-nician class, upon passing a Morse code examination operation on certain HF bands is permitted.

THE REASON FOR THE MORSE CODE EXAMINATION

Telegraphy is a method of electrical communication that the Amateur Radio Service community strongly desires to preserve. The FCC supports this objective by authorizing additional operating privileges to amateur operators who pass a Morse Code examination. Normally, to attain this skill, intense practice is required. Annually, thousands of amateur operators prove, by passing examinations, that they have acquired the skill. These examinations are prepared and administered by amateur ope-rators in the local community who volunteer their time and effort.

THE EXAMINATION PROCEDURE

The volunteer examiners (VEs) send a short message in the Morse code. The examinee must decipher a series of audible dots and dashes used in 43 different alphabetic, numeric, and punctuation characters used in the message. Usually a 10-question quiz is then administered asking questions about items contained in the mes-sage.

ACCOMMODATING A HANDICAPPED PERSON

Many handicapped persons accept and benefit from the personal challenge of passing the examination in spite of their hardships. For handicapped persons who have difficulty in proving that they can decipher messages sent in the Morse code, the VEs make exceptionally accommodative arrangements. To assist such persons, the VEs will:

- adjust the tone in frequency and volume to suit the examinee.
- administer the examination at a place convenient and com-fortable to the examinee, even at bedside.
- for a deaf person, they will send the dots and dashes to a vibrating surface or flashing light.
- write the examinee's dictation.
- where warranted, they will pause in sending the message after each sentence, each phrase, each word, or in extreme cases they will pause the exam message character-by-character to allow the examinee additional time to absorb, to interpret or even to speak out what was sent.
- or they will even allow the examinee to send the message, rather than receive it.

Should you have any questions, please contact your local volunteer examiner team, or contact one of the 14 volunteer examiner coordina-tor (VEC) organizations. For contact information for VECs, or to contact the FCC, call 888-225-5322 (weekdays), or write to FCC, 1270 Fairfield Road, Gettysburg PA 17325-7245. Fax 717-338-2696. Also see the FCC web at: <http//www.fcc.gov/wtb/amateur/>.

RENEWING, MODIFYING OR REINSTATING YOUR AMATEUR RADIO OPERATOR/PRIMARY STATION LICENSE

RENEWING YOUR AMATEUR LICENSE

The NCVEC Form 605 may also be used to renew or modify your Amateur Radio Operator/Primary Station license. License renewal may only be completed during the final 90 days prior to license expiration, or up to two years after expiration. Changes to your mailing address, name and requests for a sequential change of your station call sign appropriate for your license class may be requested at any time. This form may not be used to apply for a specific ("Vanity") station call sign.

REINSTATING YOUR AMATEUR LICENSE

This form may also be used to reinstate your Amateur Radio Operator/ Station license if it has been expired less than the two year grace period for renewal. After the two year grace period you must retake the amateur license examinations to become relicensed. You will be issued a new systematic call sign.

RENEWING OR MODIFYING YOUR LICENSE

On-line renewal: You can submit your renewal or license modifica-tions to FCC on-line via the internet/WWW at: <http://www.fcc.gov/wtb/uls>. To do so, you must first register in ULS by following the "TIN/Call Sign Registration" tab procedures, then choose the "Connecting to ULS" tab procedures and use their special dial-in to an FCC 800# modem-only access system.

Renewal by mail: If you choose to renew by mail, you can mail the "FCC Form 605" to FCC. You can obtain FCC Form 605 via the internet at <http://www.fcc.gov/formpage.html> or <ftp://ftp.fcc.gov/pub/Forms/Form605/>. It's available by fax at 202-418-0177 (request Form 000605). The FCC Forms Distribution Center will accept form orders by calling 800-418-3676. FCC Form 605 has a main form, plus a Schedule D. The main form is all that is needed for renewals. Mail FCC Form 605 to: FCC, 1270 Fairfield Rd, Gettysburg PA 17325-7245. This is a free FCC service.

The NCVEC Form 605 application can be used for a license renewal, modification or reinstatement. NCVEC Form 605 can be processed by VECs, but not all VECs provide this as a routine service. ARRL Members can submit NCVEC Form 605 to the ARRL/VEC for process-ing. ARRL Members or others can choose to submit their NCVEC Form 605 to a local VEC (check with the VEC office before forwarding), or it can be returned with a $6.00 application fee to: The W5YI Group, Inc., P.O. Box 565101, Dallas, Texas 75356 (a portion of this fee goes to the National Conference of VECs to help defray their expenses). The NCVEC Form 605 may not be returned to the FCC since it is an internal VEC form. Once again, the service provided by FCC is free.

THE FCC APPLICATION FORM 605

The FCC version of the Form 605 may not be used for applications submitted to a VE team or a VEC since it does not request information needed by the administering VEs. The FCC Form 605 may, however, be used to routinely renew or modify your license without charge. It should be sent to the FCC, 1270 Fairfield Rd., Gettysburg PA 17325-7245.

CLUB AND MILITARY RECREATION CALL SIGN ADMINISTRATORS

The NCVEC Form 605 may also be used for the processing of applications for Amateur Service club and military recreation station call signs and for the modification of RACES stations. No fee may be charged by an administrator for this service. As of March 9, 2000, FCC had not yet implemented the Call Sign Administrator System.

include a large business sized stamped, self-addressed envelope with your request.) There is a sample of those portions of an FCC Form 605 that you would complete to submit a change of address to the FCC on pages 24 and 25.

Most of the form is simple to fill out. You will need to know that the Radio Service Code for box 1 is HA for Amateur Radio. (Just remember HAM radio.) You will have to include a "Taxpayer Identification Number" on the Form. This is normally your Social Security Number. If you don't want to write your Social Security Number on this form, then you can register with the ULS as described above. Then you will receive a ULS Registration Number from the FCC, and you can use that number instead of your Social Security Number on the Form. Of course, you will have to supply your Social Security Number to register with the ULS.

The telephone number, fax number and e-mail address information is optional. The FCC will use that information to contact you in case there is a problem with your application.

Page two includes six General Certification Statements. Statement five may seem confusing. Basically, this statement means that you do not plan to install an antenna over 200 feet high, and that your permanent station location will not be in a designated wilderness area, wildlife preserve or nationally recognized scenic and recreational area.

The sixth statement indicates that you are familiar with the FCC RF Safety Rules, and that you will obey them. Subelement T0 includes exam questions and explanations about those rules.

Volunteer Examiner Program

Before you can take an FCC exam, you'll have to fill out a copy of the National Conference of Volunteer Examiner Coordinators' (NCVEC) Quick Form 605. This form is used as an application for a new license or an upgraded license. The NCVEC Quick Form 605 is only used at license exam sessions. This form includes some information that the Volunteer Examiner Coordinator's office will need to process your application with the FCC. See **Figure 4**. You should not use an NCVEC Quick Form 605 to apply for a license renewal or modification with the FCC. *Never* mail these forms to the FCC, because that will result in a rejection of the application. Likewise, an FCC Form 605 can't be used for an exam application.

All US amateur exams are administered by Volunteer Examiners who are certified by a Volunteer-Examiner Coordinator (VEC). Program. *The ARRL's FCC Rule Book* contains more details about the Volunteer-Examiner program.

To qualify for a Technician license you must pass Element 2. If you already hold a valid Novice license, then you have credit for passing Element 1 and you can earn a Technician license with Morse code credit. In that case you will be able to continue using your Novice HF privileges. In addition, your upgrade to Technician will earn you full Amateur privileges on the VHF, UHF and higher frequency bands.

The Element 2 exam consists of 35 questions taken from a pool of more than 350. The question pools for all amateur exams are maintained by a Question PoolCommittee selected by the Volunteer Examiner Coordinators.

The FCC allows Volunteer Examiners to select the questions for an amateur exam, but they must use the questions exactly as they are released by the VEC that coordinates the test session. If you attend a test session coordinated by the ARRL/VEC, your test will be designed by the ARRL/VEC or by a computer program designed by the VEC. The questions and answers will be exactly as they are printed in this book.

Finding an Exam Opportunity

To determine where and when exams will be given in your area, contact the ARRL/VEC office, or watch for announcements in the Hamfest Calendar and Coming Conventions columns in *QST*. Many local clubs sponsor exams, so they are another good source of information on exam opportunities. Upcoming exams are listed on *ARRLWeb* at: **http://www.arrl.org/arrlvec/examsearch.phtml**. Registration deadlines, and the time and location of the exams, are mentioned prominently in publicity releases about upcoming sessions.

Taking the Exam

By the time examination day rolls around, you should have already prepared yourself. This means getting your schedule, supplies and mental attitude ready. Plan your schedule so you'll get to the examination site with plenty of time to spare. There's no harm in being early. In fact, you might have time to discuss hamming with another applicant, which is a great way to calm pretest nerves. Try not to discuss the material that will be on the examination, as this may make you even more nervous. By this time, it's too late to study anyway!

What supplies will you need? First, be sure you bring your current original Amateur Radio license, if you have one. Bring a photocopy of your license, too, as well as the original and a photocopy of any Certificates of Successful Completion of Examination (CSCE) that you plan to use for exam credit. Bring along several sharpened number 2 pencils and two pens (blue or black ink). Be sure to have a good eraser. A pocket calculator may also come in handy. You may use a programmable calculator if that is the kind you have, but take it into your exam "empty" (cleared of all programs and constants in memory). Don't program equations ahead of time, because you may be asked to demonstrate that there is nothing in the calculator memory. The examining team has the right to refuse a candidate the use of any calculator that they feel may contain information for the test or could otherwise be used to cheat on the exam.

The Volunteer Examiner Team is required to check two forms of identification before you enter the test room. This includes your *original* Amateur Radio license, if you have one—not a photocopy. A photo ID of some type is best for the second form of ID, but is not required by the FCC. Other acceptable forms of identification include a driver's license, a piece of mail addressed to you or a birth certificate.

The following description of the testing procedure applies to exams coordinated by the ARRL/VEC, although many other VECs use a similar procedure.

Code Test

The code test is usually given before the written exams. If you don't plan to take the code exam, just sit quietly while the other candidates give it a try.

Before you take the code test, you'll be handed a piece of paper to copy the code as it is sent. The test will begin with about a minute of practice copy. Then comes the actual test: at least five minutes of Morse code. You are responsible for knowing the 26 letters of the alphabet, the numerals 0 through 9, the period, comma, question mark, and the procedural signals $\overline{AR}$ (+), $\overline{SK}$, $\overline{BT}$ (= or double dash) and $\overline{DN}$ (/ or fraction bar, sometimes called the "slant bar").

You may copy the entire text word for word, or just take notes on the content. At the end of the transmission, the examiner will hand you 10 questions about the text. Depending on the test format, answer the multiple-choice questions or simply fill in the blanks with your answers. (You must spell each answer exactly as it was sent.) If you get at least 7 correct, you pass! Alternatively, the exam team has the option to look at your copy sheet if you fail the 10-question exam. If you have one minute of solid copy (25 characters), the examiners can certify that you passed the test on that basis. The format of the test transmission is generally similar to one side of a normal on-the-air amateur conversation.

A sending test may not be required. The Commission has decided that if applicants can demonstrate receiving ability, they most likely can also send at that speed. But be prepared for a sending test, just in case! Subpart 97.503(a) of the FCC Rules says, "A telegraphy examination must be sufficient to prove that the examinee has the ability to send correctly by hand and to receive correctly by ear texts in the international Morse code at not less than the prescribed speed..."

Written Tests

After the code tests are administered, you'll take the written examination. The examiner will give each applicant a test booklet, an answer sheet and scratch paper. After that, you're on your own. The first thing to do is read the instructions. Be sure to sign your name every place it's called for. Do all of this at the beginning to get it out of the way.

Next, check the examination to see that all pages and questions are there. If not, report this to the examiner immediately. When filling in your answer sheet make sure your answers are marked next to the numbers that correspond to each question.

Go through the entire exam, and answer the easy questions first. Next, go back to the beginning and try the harder questions. Leave the really tough questions for last. Guessing can only help, as there is no additional penalty for answering incorrectly.

If you have to guess, do it intelligently: At first glance, you may find that you can eliminate one or more "distracters." Of the remaining responses, more than one may seem correct; only one is the best answer, however. To the applicant who is fully prepared, incorrect distracters to each question are obvious. Nothing beats preparation!

After you've finished, check the examination thoroughly. You may have read a question wrong or goofed in your arithmetic. Don't be overconfident. There's no rush, so take your time. Think, and check your answer sheet. When you feel you've done your best and can do no more, return the test booklet, answer sheet and scratch pad to the examiner.

The Volunteer-Examiner Team will grade the exam while you wait. The passing mark is 74%. (That means 26 out of 35 questions correct — or no more than 9 incorrect answers on the Element 2 exam.) You will receive a Certificate of Successful Completion of Examination (CSCE) showing all exam elements that you pass at that exam session. If you are already licensed, and you pass the exam elements required to earn a higher license class, the CSCE authorizes you to operate with your new privileges immediately. When you use these new privileges, you must sign your call sign followed by the slant mark ("/"; on voice, say "stroke" or "slant") and the letters "KT," if you are upgrading from a Novice to a Technician with code license. You only have to follow this special identification procedure until your new license is granted by the FCC, however.

If you pass only some of the exam elements required for a license, you will still receive a CSCE. That certificate shows what exam elements you passed, and is valid for 365 days. Use it as proof that you passed those exam elements so you won't have to take them over again next time you try for the license.

And Now, Let's Begin

The complete Technician question pool (Element 2) is printed in this book. Each chapter lists all the questions for a particular subelement (such as Electrical Principles — T5). A brief explanation about the correct answer is given after each question.

Table 4 shows the study guide or syllabus for the Element 2 exam as released by the Volunteer-Examiner Coordinators' Question Pool Committee in December 1999. The syllabus lists the topics to be covered by the Technician exam, and so forms the basic outline for the remainder of this book. Use the syllabus to guide your study, and to ensure that you have studied the material for all of the topics listed.

The question numbers used in the question pool refer to this syllabus. Each question number begins with a syllabus-point number (for example, T0C or T1E). The question numbers end with a two-digit number. For example, question T3B09 is the ninth question about the T3B syllabus point.

The Question Pool Committee designed the syllabus and question pool so there are the same number of points in each subelement as there are exam questions from that subelement. For example, three exam questions on the Technician exam must be from the "Radio-Wave Propagation" subelement, so there are three groups for that point. These are numbered T3A, T3B and T3C. While not a requirement of the FCC Rules, the Question Pool Committee recommends that one question be taken from each group to make the best possible license exams.

Good luck with your studies!

Table 4
Technician Class (Element 2) Syllabus

(Required for all operator licenses.)

Subelement T1 — Commission's Rules

[9 Exam Questions — 9 Groups]

T1A Basis and purpose of amateur service and definitions; Station / Operator license; classes of US amateur licenses, including basic differences; privileges of the various license classes; term of licenses; grace periods; modifications of licenses; current mailing address on file with FCC

T1B Frequency privileges authorized to the Technician control operator (VHF/UHF and HF)

T1C Emission privileges authorized to the Technician control operator (VHF/UHF and HF)

T1D Responsibility of licensee; station control; control operator requirements; station identification; points of communication and operation; business communications

T1E Third-party communication; authorized and prohibited transmissions; permissible one-way communication

T1F Frequency selection and sharing; transmitter power; digital communications

T1G Satellite and space communications; false signals or unidentified communications; malicious interference

T1H Correct language; phonetics; beacons; radio control of model craft and vehicles

T1I Emergency communications; broadcasting; indecent and obscene language

Subelement T2 — Operating Procedures

[5 Exam Questions — 5 Groups]

T2A Preparing to transmit; choosing a frequency for tune-up; operating or emergencies; Morse code; repeater operations and autopatch

T2B Definition and proper use; courteous operation; repeater frequency coordination; Morse code

T2C Simplex operations; RST signal reporting; choice of equipment for desired communications; communications modes including amateur television (ATV), packet radio; Q signals, procedural signals and abbreviations

T2D Distress calling and emergency drills and communications — operations and equipment; Radio Amateur Civil Emergency Service (RACES)

T2E Voice communications and phonetics; SSB/CW weak signal operations; radioteleprinting; packet; special operations

Subelement T3 — Radio-Wave Propagation

[3 Exam Questions — 3 Groups]

T3A Line of sight; reflection of VHF/UHF signals

T3B Tropospheric ducting or bending; amateur satellite and EME operations

T3C Ionospheric propagation, causes and variation; maximum usable frequency; Sporadic-E propagation; ground wave, HF propagation characteristics; sunspots and the sunspot cycle

Subelement T4 — Amateur Radio Practices

[4 Exam Questions — 4 Groups]

T4A Lightning protection and station grounding; safety interlocks, antenna installation safety procedures; dummy antennas

T4B Electrical wiring, including switch location, dangerous voltages and currents; SWR meaning and measurements; SWR meters

T4C Meters and their placement in circuits, including volt, amp, multi, peak-reading and RF watt; ratings of fuses and switches

T4D RFI and its complications, resolution and responsibility

Subelement T5 — Electrical Principles

[3 Exam Questions — 3 Groups]

T5A Metric prefixes, e.g. pico, nano, micro, milli, centi, kilo, mega, giga; concepts, units and measurement of current, voltage; concept of conductor and insulator; concept of open and short circuits

T5B Concepts, units and calculation of resistance, inductance and capacitance values in series and parallel circuits

T5C Ohm's Law (any calculations will be kept to a very low level - no fractions or decimals) and the concepts of energy and power; concepts of frequency, including AC vs. DC, frequency units, and wavelength

Subelement T6 — Circuit Components

[2 Exam Questions — 2 Groups]

T6A Electrical function and/or schematic representation of resistor, switch, fuse, or battery; resistor construction types, variable and fixed, color code, power ratings, schematic symbols

T6B Electrical function and/or schematic representation of a ground, antenna, inductor, capacitor, transistor, integrated circuit; construction of variable and fixed inductors and capacitors; factors affecting inductance and capacitance

Subelement T7 — Practical Circuits

[2 Exam Questions — 2 Groups]

T7A Functional layout of station components including transmitter, transceiver, receiver, power supply, antenna, antenna switch, antenna feed line, impedance-matching device, SWR meter; station layout and accessories for radiotelephone, radioteleprinter (RTTY) or packet

T7B Transmitter and receiver block diagrams; purpose and operation of low-pass, high-pass and band-pass filters

Subelement T8 — Signals and Emissions

[2 Exam Questions — 2 Groups]

T8A RF carrier, definition and typical bandwidths; harmonics and unwanted signals; chirp; superimposed hum; equipment and adjustments to help reduce interference to others

T8B Concepts and types of modulation: CW, phone, RTTY and data emission types; FM deviation

Subelement T9 — Antennas and Feed Lines

[2 Exam Questions — 2 Groups]

T9A Wavelength vs. antenna length; 1/2 wavelength dipole and 1/4 wavelength vertical antennas; multiband antennas

T9B Parasitic beam directional antennas; polarization, impedance matching and SWR, feed lines, balanced vs. unbalanced (including baluns)

Subelement T0 — RF Safety

[3 Exam Questions — 3 Groups]

T0A RF safety fundamentals, terms and definitions

T0B RF safety rules and guidelines

T0C Routine station evaluation (Practical applications for VHF/UHF and above operations)

Commission's Rules

Your Technician (Element 2) exam will consist of 35 questions, taken from the Technician question pool, as prepared by the Volunteer Examiner Coordinators' Question Pool Committee. A certain number of questions are taken from each of the 10 subelements. There will be 9 questions from the Commission's Rules subelement printed in this chapter. These questions are divided into 9 groups, labeled T1A through T1I.

After most of the explanations in this chapter you will see a reference to Part 97 of the FCC Rules set inside square brackets, like [97.3]. This tells you where to look for the exact wording of the Rules as they relate to that question. For a complete copy of Part 97, along with simple explanations of the Rules governing Amateur Radio, see *The FCC Rule Book* published by the ARRL.

T1A Basis and purpose of amateur service and definitions; Station/Operator licenses; classes of US amateur licenses, including basic differences; privileges of the various license classes; term of licenses; grace periods; modifications of licenses; current mailing address on file with FCC

T1A01 Who makes and enforces the rules and regulations of the amateur service in the US?

A. The Congress of the United States
B. The Federal Communications Commission (FCC)
C. The Volunteer Examiner Coordinators (VECs)
D. The Federal Bureau of Investigation (FBI)

B　The Federal Communications Commission (FCC) is responsible for creating the rules to govern the Amateur Radio Service in the US. The FCC also enforces those rules. [97]

T1A02 What are two of the five purposes for the amateur service?

A. To protect historical radio data, and help the public understand radio history

B. To help foreign countries improve communication and technical skills, and encourage visits from foreign hams

C. To modernize radio schematic drawings, and increase the pool of electrical drafting people

D. To increase the number of trained radio operators and electronics experts, and improve international goodwill

D The Rules and Regulations of Part 97 are designed to provide an Amateur Radio Service that has a fundamental purpose described by five principles. These principles recognize the value of the amateur service in providing emergency communication; advancing the radio art; gaining technical and operating skills; providing **trained operators, technicians and electronics experts**; and **enhancing international goodwill**. [97.1]

T1A03 What is the definition of an amateur station?

A. A station in a public radio service used for radiocommunications

B. A station using radiocommunications for a commercial purpose

C. A station using equipment for training new broadcast operators and technicians

D. A station in the Amateur Radio service used for radiocommunications

D An amateur station is any collection of equipment necessary to **communicate with other stations** that are part of the **Amateur Radio Service**. [97.3 (a) (5)]

T1A04 What is the definition of a control operator of an amateur station?

A. Anyone who operates the controls of the station

B. Anyone who is responsible for the station's equipment

C. Any licensed amateur operator who is responsible for the station's transmissions

D. The amateur operator with the highest class of license who is near the controls of the station

C A control operator is responsible for the **signals** being **transmitted from an Amateur Radio station**, and must have an amateur license. [97.3 (a) (12)]

T1A05 Which of the following is required before you can operate an amateur station in the US?

A. You must hold an FCC operator's training permit for a licensed radio station

B. You must submit an FCC Form 605 together with a license examination fee

C. The FCC must grant you an amateur operator/primary station license

D. The FCC must issue you a Certificate of Successful Completion of Amateur Training

C   You must have an Amateur Radio license to operate an amateur station in the US. The FCC calls this an **amateur operator/primary station license** because one license authorizes you to be a control operator and to have your own Amateur Radio station. Your amateur license has been granted as soon as the records appear in the FCC database. [97.5 (a), (b) (1)]

T1A06 What must happen before you are allowed to operate an amateur station?

A. The FCC database must show that you have been granted an amateur license

B. You must have written authorization from the FCC

C. You must have written authorization from a Volunteer Examiner Coordinator

D. You must have a copy of the FCC Rules, Part 97, at your station location

A   Before you can be the control operator of an amateur station the FCC must grant you an amateur license. You don't have to wait for that license to arrive in the mail, however. Your license is valid as soon as the **license information appears in the FCC database**. [97.5 (a), 97.7 (a)]

T1A07 What are the US amateur operator licenses that a new amateur might earn?

A. Novice, Technician, General, Advanced

B. Technician, Technician Plus, General, Advanced

C. Novice, Technician, General, Advanced

D. Technician, Technician with Morse code, General, Amateur Extra

D   The classes of US amateur operator licenses that a new amateur can earn are **Technician, Technician with Morse code, General and Amateur Extra**. [97.9 (a)].

T1A08 How soon after you pass the elements required for your first Amateur Radio license may you transmit?

A. Immediately
B. 30 days after the test date
C. As soon as the FCC grants you a license
D. As soon as you receive your license from the FCC

C   **The FCC must grant you an amateur license** before you may transmit as the control operator of an amateur station . You don't have to wait for that license to arrive in the mail, however. Your license is valid as soon as the license information appears in the FCC database. [97.5 (a), 97.7 (a)]

T1A09 How soon before the expiration date of your license should you send the FCC a completed Form 605 or file with the Universal Licensing System on the World Wide Web for a renewal?

A. No more than 90 days
B. No more than 30 days
C. Within 6 to 9 months
D. Within 6 months to a year

A   The FCC recommends that you submit a Form 605 for renewal of your license between 60 and 90 days before the current license expires. Be sure it is **no more than 90 days** before the expiration date when you submit your renewal application to the FCC. You can send your Form 605 to a Volunteer Examiner Coordinator and ask them to file your renewal electronically. The ARRL/VEC will file the forms electronically for ARRL members free of charge. You can also use the FCC's Universal Licensing System (ULS) to file your renewal. Go to **http://www.fcc.gov/wtb/uls** and click on the "TIN/Call Sign Registration" button. Follow the directions to register with the ULS. Next click on the "Connecting to ULS" button and follow the directions given there to connect to the FCC's ULS database. [97.21 (a) (3)]

T1A10 What is the normal term for which a new amateur station license is granted?

A. 5 years
B. 7 years
C. 10 years
D. For the lifetime of the licensee

C   The FCC issues new amateur licenses for a **10-year** term. [97.25]

T1A11 What is the "grace period" during which the FCC will renew an expired 10-year license?

- A. 2 years
- B. 5 years
- C. 10 years
- D. There is no grace period

A   If you forget to renew your license, you have up to **two years** to apply for a new license. Apply for your license renewal by completing an FCC Form 605 and sending it to the FCC in Gettysburg, PA. Your license is not valid during that two-year grace period, and you may not operate your station until you renew your license. The FCC will renew your license without requiring you to pass the exams again during the grace period. After the two-year grace period you will have to pass the exam again. You can also have a VEC file your application electronically or use the FCC Universal Licensing System during the grace period. See the explanation to question T1A09 for details. [97.21 (b)]

T1A12 What is one way you may notify the FCC if your mailing address changes?

- A. Fill out an FCC Form 605 using your new address, attach a copy of your license, and mail it to your local FCC Field Office
- B. Fill out an FCC Form 605 using your new address, attach a copy of your license, and mail it to the FCC office in Gettysburg, PA
- C. Call your local FCC Field Office and give them your new address over the phone or e-mail this information to the local Field Office
- D. Call the FCC office in Gettysburg, PA, and give them your new address over the phone or e-mail this information to the FCC

B   Use an **FCC Form 605** for any license renewal or modification. Check the "CHANGE my mailing address on my license to my new address" box in Section 4 of the form. **Mail** the completed form **to the FCC in Gettysburg, PA**. You can also a VEC to file your renewal electronically, or use the ULS, but those options are not included as the answer choices to this question. [97.21 (a) (1), 97.23]

FCC 605
Approved by OMB
Main Form

Quick-Form Application for Authorization in the Ship, Aircraft,

Amateur, Restricted and Commercial Operator, and the
General Mobile Radio Services

3060 - 0850
See instructions for
public burden estimate

---

1) Radio Service Code: **HA**

Application Purpose (Select only one) **(MD)**

2)	NE - New	RO - Renewal Only	WD - Withdrawal of Application
	MD - Modification	RM - Renewal/Modification	DU - Duplicate License
	AM - Amendment	CA - Cancellation of License	AU - Administrative Update

3)	If this request is for a **D**evelopmental License or **S**TA (Special Temporary Authorization) enter the appropriate code and attach the required exhibit as described in the instructions. Otherwise enter **N** (Not Applicable).	**(N)D S N/A**
4)	If this request is for an Amendment or Withdrawal of Application, enter the file number of the pending application currently on file with the FCC.	File Number
5)	If this request is for a Modification, Renewal Only, Renewal/Modification, Cancellation of License, Duplicate License, or Administrative Update, enter the call sign of the existing FCC license.	Call Sign **WR1B**
6)	If this request is for a New, Amendment, Renewal Only, or Renewal/Modification, enter the requested authorization expiration date (this item is optional).	MM DD
7)	Does this filing request a Waiver of the Commission's rules? If 'Y', attach the required showing as described in the instructions.	**(N)Yes No**
8)	Are attachments (other than associated schedules) being filed with this application?	**(N)Yes No**

Applicant Information

9a) Taxpayer Identification Number: **123 - 45 - 6789**		9b) SGIN:

10) Applicant/Licensee is a(n): (**I**) Individual Unincorporated Association Trust Government Entity Joint Venture
    Corporation Limited Liability Corporation Partnership Consortium

11) First Name (if individual): **Larry**	MI: **D**	Last Name: **Wolfgang**	Suffix:

12) Entity Name (if other than individual):

13) Attention To:

14) P.O. Box:	And/Or	15) Street Address: **225 Main Street**

16) City: **Newington**	17) State: **CT**	18) Zip: **06111**	19) Country: **USA**

20) Telephone Number: **860-594-0200**	21) FAX:

22) E-Mail Address: **wr1b@arrl.net**

FCC 605- Main Form
July 1999 - Page 1

Fee Status

23) Is the applicant exempt from FCC application fees?	(*N*)Yes No	
24) Is the applicant exempt from FCC regulatory fees?	(*N*)Yes No	

General Certification Statements

1) The Applicant waives any claim to the use of any particular frequency or of the electromagnetic spectrum as against the regulatory power of the United States because of the previous use of the same, whether by license or otherwise, and requests an authorization in accordance with this application.

2) The applicant certifies that all statements made in this application and in the exhibits, attachments, or documents incorporated by reference are material, are part of this application, and are true, complete, correct, and made in good faith.

3) Neither the Applicant nor any member thereof is a foreign government or a representative thereof.

4) The applicant certifies that neither the applicant nor any other party to the application is subject to a denial of Federal benefits pursuant to Section 5301 of the Anti-Drug Abuse Act of 1988, 21 U.S.C. § 862, because of a conviction for possession or distribution of a controlled substance. **This certification does not apply to applications filed in services exempted under Section 1.2002(c) of the rules, 47 CFR § 1.2002(c).** See Section 1.2002(b) of the rules, 47 CFR § 1.2002(b), for the definition of "party to the application" as used in this certification.

5) Amateur or GMRS Applicant certifies that the construction of the station would NOT be an action which is likely to have a significant environmental effect (see the Commission's Rules 47 CFR Sections 1.1301-1.1319 and Section 97.13(a).

6) Amateur Applicant certifies that they have READ and WILL COMPLY WITH Section 97.13(c) of the Commission's Rules regarding RADIOFREQUENCY (RF) RADIATION SAFETY and the amateur service section of OST/OET Bulletin Number 65.

Certification Statements For GMRS Applicants

1) Applicant certifies that he or she is claiming eligibility under Rule Section 95.5 of the Commission's Rules.

2) Applicant certifies that he or she is at least 18 years of age.

3) Applicant certifies that he or she will comply with the requirement that use of frequencies 462.650, 467.650, 462.700 and 467.700 MHz is not permitted near the Canadian border North of Line A and East of Line C. These frequencies are used throughout Canada and harmful interference is anticipated.

Signature

25) Typed or Printed Name of Party Authorized to Sign

First Name: Larry	MI: D	Last Name: Wolfgang	Suffix:

26) Title:

Signature: *Larry D. Wolfgang*	27) Date: 4/4/00

Failure To Sign This Application May Result In Dismissal Of The Application And Forfeiture Of Any Fees Paid

WILLFUL FALSE STATEMENTS MADE ON THIS FORM OR ANY ATTACHMENTS ARE PUNISHABLE BY FINE AND/OR IMPRISONMENT (U.S. Code, Title 18, Section 1001) AND/OR REVOCATION OF ANY STATION LICENSE OR CONSTRUCTION PERMIT (U.S. Code, Title 47, Section 312(a)(1)), AND/OR FORFEITURE (U.S. Code, Title 47, Section 503).

T1B Frequency privileges authorized to the Technician control operator (VHF/UHF and HF)

T1B01 What are the frequency limits of the 6-meter band in ITU Region 2?

    A. 52.0 - 54.5 MHz
    B. 50.0 - 54.0 MHz
    C. 50.1 - 52.1 MHz
    D. 50.0 - 56.0 MHz

B   Technicians may operate on the entire 6-meter band, **50.0 - 54.0 MHz**. ITU Region 2 primarily covers North and South America. [97.301 (a)]

T1B02 What are the frequency limits of the 2-meter band in ITU Region 2?

    A. 145.0 - 150.5 MHz
    B. 144.0 - 148.0 MHz
    C. 144.1 - 146.5 MHz
    D. 144.0 - 146.0 MHz

B   Technicians may operate on the entire 2-meter band, **144.0 - 148.0 MHz**. ITU Region 2 primarily covers North and South America. [97.301 (a)]

T1B03 What are the frequency limits of the 1.25-meter band in ITU Region 2?

    A. 225.0 - 230.5 MHz
    B. 222.0 - 225.0 MHz
    C. 224.1 - 225.1 MHz
    D. 220.0 - 226.0 MHz

B   Technicians may operate on the entire 1.25-meter band, **222.0 - 225.0 MHz**. ITU Region 2 primarily covers North and South America. [97.301 (a)]

T1B04 What are the frequency limits of the 70-centimeter band in ITU Region 2?

    A. 430-0 - 440.0 MHz
    B. 430.0 - 450.0 MHz
    C. 420.0 - 450.0 MHz
    D. 432.0 - 435.0 MHz

C   Technicians may operate on the entire 70-cm band, **420.0 - 450.0 MHz**. ITU Region 2 primarily covers North and South America. [97.301 (a)]

T1B05 What are the frequency limits of the 33-centimeter band in ITU Region 2?

A. 903 - 927 MHz
B. 905 - 925 MHz
C. 900 - 930 MHz
D. 902 - 928 MHz

D   Technicians may operate on the entire 33-cm band, **902 - 928 MHz**. ITU Region 2 primarily covers North and South America. [97.301 (a)]

T1B06 What are the frequency limits of the 23-centimeter band?

A. 1260 - 1270 MHz
B. 1240 - 1300 MHz
C. 1270 - 1295 MHz
D. 1240 - 1246 MHz

B   Technicians may operate on the entire 23-cm band, **1240 - 1300 MHz**. ITU Region 2 primarily covers North and South America. [97.301 (a)]

T1B07 What are the frequency limits of the 13-centimeter band in ITU Region 2?

A. 2300 - 2310 MHz and 2390 - 2450 MHz
B. 2300 - 2350 MHz and 2400 - 2450 MHz
C. 2350 - 2380 MHz and 2390 - 2450 MHz
D. 2300 - 2350 MHz and 2380 - 2450 MHz

A   Technicians may operate on the entire 13-cm band, **2300 - 2310 and 2390 - 2450 MHz**. ITU Region 2 primarily covers North and South America. [97.301 (a)]

T1B08 What are the frequency limits of the 80-meter band for Technician class licensees who have passed a Morse code exam?

A. 3500 - 4000 kHz
B. 3675 - 3725 kHz
C. 7100 - 7150 kHz
D. 7000 - 7300 kHz

B   Technicians with code credit may operate on the 80-meter band from **3675 kHz - 3725 kHz**. [97.301 (e)]

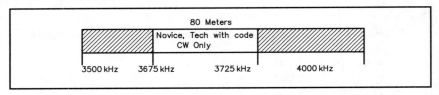

T1B09 What are the frequency limits of the 40-meter band in ITU Region 2 for Technician class licensees who have passed a Morse code exam?

A. 3500 - 4000 kHz
B. 3700 - 3750 kHz
C. 7100 - 7150 kHz
D. 7000 - 7300 kHz

C   Technicians with code credit may operate on the 40-meter band from **7100 kHz - 7150 kHz**. ITU Region 2 consists primarily of North and South America. [97.301 (e)]

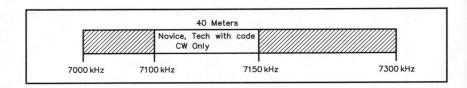

T1B10 What are the frequency limits of the 15-meter band for Technician class licensees who have passed a Morse code exam?

A. 21.100 - 21.200 MHz
B. 21.000 - 21.450 MHz
C. 28.000 - 29.700 MHz
D. 28.100 - 28.200 MHz

A   Technicians with code credit may operate on the 15-meter band from **21.100 MHz - 21.200 MHz**. [97.301 (e)].

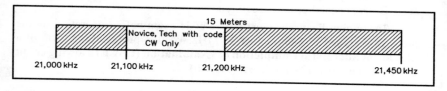

T1B11 What are the frequency limits of the 10-meter band for Technician class licensees who have passed a Morse code exam?

A. 28.000 - 28.500 MHz
B. 28.100 - 29.500 MHz
C. 28.100 - 28.500 MHz
D. 29.100 - 29.500 MHz

C   Technicians with code credit may operate on the **28.100 - 28.500 MHz** portion of the 10-meter band. [97.301 (e)]

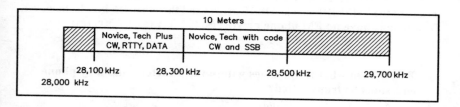

T1B12 If you are a Technician licensee who has passed a Morse code exam, what is the one document you can use to prove that you are authorized to use certain amateur frequencies below 30 MHz?

A. A certificate from the FCC showing that you have notified them that you will be using the HF bands
B. A certificate showing that you have attended a class in HF communications
C. A Certificate of Successful Completion of Examination showing that you have passed a Morse code exam
D. No special proof is required

C   Technician licensees must pass a code exam before operating on the frequencies assigned to them below 30 MHz. Written proof, in the form of a **Certificate of Successful Completion of Examination (CSCE)** showing credit for a **Morse code exam**, is required. [97.9 (b)]

T1C Emission privileges authorized to the Technician control operator (VHF/UHF and HF)

T1C01 On what HF band may a Technician licensee use FM phone emission?

A. 10 meters
B. 15 meters
C. 75 meters
D. None

D   With your Technician (or higher) license you can operate FM phone on any of the VHF/UHF bands. On 10 meters and the other **HF bands**, however, **Technicians have no FM phone privileges**. [97.301 (a), 97.305 (c)]

T1C02 On what frequencies within the 6-meter band may phone emissions be transmitted?

A. 50.0 - 54.0 MHz only
B. 50.1 - 54.0 MHz only
C. 51.0 - 54.0 MHz only
D. 52.0 - 54.0 MHz only

B   Technician and higher licensees may transmit phone (voice) emissions from **50.1 to 54.0 MHz** on the 6-meter band. 50.0 to 50.1 MHz is reserved for CW operation. [97.305 (c)]

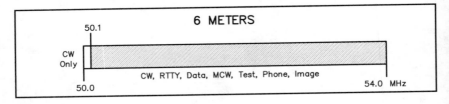

T1C03 On what frequencies within the 2-meter band may image emissions be transmitted?

    A. 144.1 - 148.0 MHz only
    B. 146.0 - 148.0 MHz only
    C. 144.0 - 148.0 MHz only
    D. 146.0 - 147.0 MHz only

    A   Technician and higher class licensees may transmit image (TV) emissions from **144.1 to 148.0 MHz** on the 2-meter band. 144.0 to 144.1 MHz is reserved for CW operation. [97.305 (c)]

T1C04 What frequencies within the 2-meter band are reserved exclusively for CW operations?

    A. 146 - 147 MHz
    B. 146.0 - 146.1 MHz
    C. 145 - 148 MHz
    D. 144.0 - 144.1 MHz

    D   The frequency range from **144.0 to 144.1 MHz** is reserved for the exclusive use of CW. [97.301 (a), 97.305 (c)]

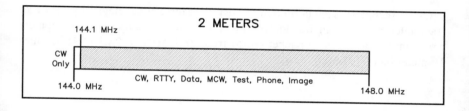

T1C05 What emission types are Technician control operators who have passed a Morse code exam allowed to use in the 80-meter band?

    A. CW only
    B. Data only
    C. RTTY only
    D. Phone only

    A   Technicians are **only** permitted to use **CW** (Morse code telegraphy) on the 80-meter band. [97.305, 97.307 (f) (9)]

T1C06 What emission types are Technician control operators who have passed a Morse code exam allowed to use from 7100 to 7150 kHz in ITU Region 2?

A. CW and data
B. Phone
C. Data only
D. CW only

D   Technicians are **only** permitted to use **CW** (Morse code telegraphy) on the frequencies from 7100 kHz to 7150 kHz in the 40-meter band. ITU Region 2 primarily covers North and South America. [97.305, 97.307 (f) (9)]

T1C07 What emission types are Technician control operators who have passed a Morse code exam allowed to use on frequencies from 28.1 to 28.3 MHz?

A. All authorized amateur emission privileges
B. Data or phone
C. CW, RTTY and data
D. CW and phone

C   Technicians are permitted to use **CW** (Morse code telegraphy), **RTTY** (radioteletype, including Baudot RTTY and AMTOR) and **data** (digital computer communications, such as packet radio) from 28.1 MHz to 28.3 MHz, on the 10-meter band. [97.305]

T1C08 What emission types are Technician control operators who have passed a Morse code exam allowed to use on frequencies from 28.3 to 28.5 MHz?

A. All authorized amateur emission privileges
B. CW and data
C. CW and single-sideband phone
D. Data and phone

C   Technicians are permitted to use **CW** (Morse code telegraphy) and **single-sideband phone** (SSB) from 28.3 MHz to 28.5 MHz on the 10-meter band. [97.305, 97.307 (f) (10)]

T1C09 What emission types are Technician control operators allowed to use on the amateur 1.25-meter band in ITU Region 2?

A. Only CW and phone
B. Only CW and data
C. Only data and phone
D. All amateur emission privileges authorized for use on the band

D   Technicians in ITU Region 2 may use **any emission type** that is permitted for any amateur operator in the 1.25-meter band, which extends from 222 MHz to 225 MHz. ITU Region 2 primarily covers North and South America. [97.305]

T1C10 What emission types are Technician control operators allowed to use on the amateur 23-centimeter band?

A. Only data and phone
B. Only CW and data
C. Only CW and phone
D. All amateur emission privileges authorized for use on the band

D   Technicians may use **any emission type** that is permitted for any amateur operator in the 23-cm band, which extends from 1240 MHz to 1300 MHz. [97.305]

T1C11 On what frequencies within the 70-centimeter band in ITU Region 2 may image emissions be transmitted?

A. 420.0 - 420.1 MHz only
B. 430.0 - 440.0 MHz only
C. 420.0 - 450.0 MHz only
D. 440.0 - 450.0 MHz only

C   Image emissions may be used on the entire 70 cm band, which extends from **420.0 - 450.0 MHz**. [97.305]

T1D   Responsibility of licensee; station control; control operator requirements; station identification; points of communication and operation; business communications

T1D01 What is the control point of an amateur station?

A. The on/off switch of the transmitter
B. The input/output port of a packet controller
C. The variable frequency oscillator of a transmitter
D. The location at which the control operator function is performed

D   The control point of an amateur station is the **location** where the **control operator** controls the station. The control point for most Amateur Radio stations is at the radio itself, although some stations are controlled in other ways. [97.3 (a) (13)]

T1D02 Who is responsible for the proper operation of an amateur station?

A. Only the control operator
B. Only the station licensee
C. Both the control operator and the station licensee
D. The person who owns the station equipment

C   The **station licensee** is always responsible for the proper operation of an amateur station, but the **control operator** is also responsible. If the control operator is someone other than the station licensee, **both share the responsibility** for the proper operation of the station. [97.103 (a)]

T1D03 What is your responsibility as a station licensee?

A. You must allow another amateur to operate your station upon request
B. You must be present whenever the station is operated
C. You must notify the FCC if another amateur acts as the control operator
D. You are responsible for the proper operation of the station in accordance with the FCC rules

D   As an amateur station licensee, **you must ensure that your station is always operated according to the FCC Rules**. [97.103 (a)]

T1D04 Who may be the control operator of an amateur station?

A. Any person over 21 years of age
B. Any person over 21 years of age with a General class license or higher
C. Any licensed amateur chosen by the station licensee
D. Any licensed amateur with a Technician class license or higher

C   You must **have an amateur license** to be the control operator of an amateur station. You must also **have the permission of the station licensee**, but that is only a concern if you want to operate someone else's station. It should be pretty easy to give yourself permission to operate your station! [97.103 (b)]

T1D05 If you are the control operator at the station of another amateur who has a higher class license than yours, what operating privileges are you allowed?

- A. Any privileges allowed by the higher license
- B. Only the privileges allowed by your license
- C. All the emission privileges of the higher license, but only the frequency privileges of your license
- D. All the frequency privileges of the higher license, but only the emission privileges of your license

B   Your amateur license allows you to have an Amateur Radio station, and also to be the control operator of a station. Your **license class determines the control-operator privileges** you may use. When you are the control operator, you may only use the privileges allowed by your license class, even if the station licensee has a higher-class operator's license. [97.105 (b)]

T1D06 When an amateur station is transmitting, where must its control operator be?

- A. At the station's control point
- B. Anywhere in the same building as the transmitter
- C. At the station's entrance, to control entry to the room
- D. Anywhere within 50 km of the station location

A   The control operator must be present **at the control point** of a station whenever the station is transmitting. [97.109 (b)]

T1D07 How often must an amateur station be identified?

- A. At the beginning of a contact and at least every ten minutes after that
- B. At least once during each transmission
- C. At least every ten minutes during and at the end of a contact
- D. At the beginning and end of each transmission

C   You must identify your station **at least once every ten minutes during a contact, and** you must also identify your station **at the end of the contact**. The FCC Rules don't specifically require you to identify your station at the beginning of a contact. It is a good idea to give your call sign when you start a contact, however, so the other station will know who you are. [97.119 (a)]

T1D08 What identification, if any, is required when two amateur stations begin communications?

    A. No identification is required
    B. One of the stations must give both stations' call signs
    C. Each station must transmit its own call sign
    D. Both stations must transmit both call signs

    A   The **FCC Rules do not require you to give your call sign at the beginning of a contact.** It is a good idea to give your call sign so the other station knows who you are, however. [97.119 (a)]

T1D09 What identification, if any, is required when two amateur stations end communications?

    A. No identification is required
    B. One of the stations must transmit both stations' call signs
    C. Each station must transmit its own call sign
    D. Both stations must transmit both call signs

    C   Always **identify your station by transmitting your call sign at the end of a communication**. Notice that you don't have to give the other station's call sign, but even if you do, the other station still must transmit its own call sign. [97.119 (a)]

T1D10 What is the longest period of time an amateur station can operate without transmitting its call sign?

    A. 5 minutes
    B. 10 minutes
    C. 15 minutes
    D. 30 minutes

    B   You must identify your station at least every **ten minutes**, so you should never transmit for more than ten minutes without giving your call sign. [97.119 (a)]

T1D11 What emission type may always be used for station identification, regardless of the transmitting frequency?

    A. CW
    B. RTTY
    C. MCW
    D. Phone

    A   **CW** (Morse code) is the one emission type that can be used on all amateur frequencies. You can always use Morse code to give your station identification, regardless of your operating frequency or mode. [97.305 (a)]

T1D12 If you are a Technician licensee with a Certificate of Successful Completion of Examination (CSCE) for a Morse code exam, how should you identify your station when transmitting on the 10 meter band?

A. You must give your call sign followed by the words "plus plus"
B. You must give your call sign followed by the words "temporary plus"
C. No special form of identification is needed
D. You must give your call sign and the location of the VE examination where you obtained the CSCE

C   If you hold a Technician license and then pass the 5 word-per-minute Morse code exam, you can begin operating on the HF bands with your new license privileges immediately. You will receive a Certificate of Successful Completion of Examination (CSCE) at the exam session, and that paper is your authority to operate with the new privileges. In this one case, **you don't have to use any form of special identification**. Just give your call sign to identify your station. [97.119 (f)]

T1E   Third-party communication; authorized and prohibited transmissions; permissible one-way communication

T1E01 What kind of payment is allowed for third-party messages sent by an amateur station?

A. Any amount agreed upon in advance
B. Donation of repairs to amateur equipment
C. Donation of amateur equipment
D. No payment of any kind is allowed

D   Amateurs often transmit messages for other people; these are called *third-party messages* or third-party communications. Amateur operators **may not accept any type of payment** for these communications because the Amateur Radio Service is a noncommercial, volunteer communication service. [97.113 (a) (2)]

T1E02 What is the definition of third-party communications?

A. A message sent between two amateur stations for someone else
B. Public service communications for a political party
C. Any messages sent by amateur stations
D. A three-minute transmission to another amateur

A   Any **message sent between two amateur stations for another person** is *called third-party communications*. (You are the *first* party and the amateur you send the message to — or receive it from — is the *second* party, so the person for whom the message is sent is the *third* party.) [97.3 (a) (46)]

T1E03 What is a "third party" in amateur communications?

    A. An amateur station that breaks in to talk

    B. A person who is sent a message by amateur communications other than a control operator who handles the message

    C. A shortwave listener who monitors amateur communications

    D. An unlicensed control operator

B   When a **message is sent between two amateur stations for another person, the other person is called a third party**. (You are the *first* party and the amateur you send the message to — or receive it from — is the *second* party, so the person for whom the message is sent is the *third* party.) [97.3 (a) (46)]

T1E04 When are third-party messages allowed to be sent to a foreign country?

    A. When sent by agreement of both control operators

    B. When the third party speaks to a relative

    C. They are not allowed under any circumstances

    D. When the US has a third-party agreement with the foreign country or the third party is qualified to be a control operator

D   Amateur Radio operators often exchange messages for other people. Such messages are called *third-party traffic* or third-party messages. Many countries do not permit their amateurs to exchange such messages with other countries, however. Before you transmit a message for someone else to a foreign country, check to be sure that the US has a **third-party agreement** with that country. If the other person (the *third party*) is a licensed amateur **qualified** to be the **control operator**, then the exchange is permitted even if no third-party agreement exists. [97.115 (a) (2)]

T1E05 If you let an unlicensed third party use your amateur station, what must you do at your station's control point?

    A. You must continuously monitor and supervise the third-party's participation

    B. You must monitor and supervise the communication only if contacts are made in countries that have no third-party communications agreement with the US

    C. You must monitor and supervise the communication only if contacts are made on frequencies below 30 MHz

    D. You must key the transmitter and make the station identification

A   You may allow an unlicensed person to talk into a mike, type on a keyboard or send Morse code over your radio. This type of communication is called *third-party communication*. When an unlicensed third party uses your station, you must **continuously monitor** the communication and **supervise** the third party to ensure no rules are violated. [97.115 (b) (1)]

International Third-Party Traffic — Proceed With Caution

Occasionally, DX stations may ask you to pass a third-party message to a friend or relative in the States. This is all right as long as the US has signed an official third-party traffic agreement with that particular country, or the third party is a licensed amateur. The traffic must be noncommercial and of a personal, unimportant nature. During an emergency, the US State Department will often work out a special temporary agreement with the country involved. But in normal times, never handle traffic without first making sure it is legally permitted.

US Amateurs May Handle Third-Party Traffic With:

V2	Antigua/Barbuda	6Y	Jamaica
LU	Argentina	JY	Jordan
VK	Australia	EL	Liberia
V3	Belize	V7	Marshall Islands
CP	Bolivia	XE	Mexico
T9	Bosnia-Herzegovina	V6	Micronesia, Federated
PY	Brazil		States of
VE	Canada	YN	Nicaragua
CE	Chile	HP	Panama
HK	Colombia	ZP	Paraguay
D6	Comoros (Federal	OA	Peru
	Islamic Republic of)	DU	Philippines
TI	CostaRica	VR6	Pitcairn Island*
CO	Cuba	V4	St. Christopher/Nevis
HI	DominicanRepublic	J6	St. Lucia
J7	Dominica	J8	St. Vincent and the
HC	Ecuador		Grenadines
YS	El Salvador	9L	Sierra Leone
C5	Gambia, The	ZS	South Africa
9G	Ghana	3DA	Swaziland
J3	Grenada	9Y	Trinidad/Tobago
TG	Guatemala	TA	Turkey
8R	Guyana	GB	United Kingdom**
HH	Haiti	CX	Uruguay
HR	Honduras	YV	Venezuela
4X	Israel	4U1ITU	ITU Geneva
		4U1VIC	VIC Vienna

Notes:

*Since 1970, there has been an informal agreement between the United Kingdom and the US, permitting Pitcairn and US amateurs to exchange messages concerning medical emergencies, urgent need for equipment or supplies, and private or personal matters of island residents.
**Limited to special-event stations with callsign prefix GB (GB3 excluded).

US licensed amateurs may operate in the following US territories under their FCC license: The Northern Marianas Islands, Guam, Johnston Island, Midway Island, Kure Island, American Samoa, Wake Island, Wilkes Island, Peale Island, The Commonwealth of Puerto Rico and the US Virgin Islands.

Please note that the Region 2 Division of the International Amateur Radio Union (IARU) has recommended that international traffic on the 20 and 15-meter bands be conducted on the following frequencies:

14.100-14.150 MHz   21.150-21.200 MHz
14.250-14.350 MHz   21.300-21.450 MHz

The IARU is the alliance of Amateur Radio societies from around the world; Region 2 comprises member-societies in North, South and Central America, and the Caribbean.

Note: At the end of an exchange of third-party traffic with a station located in a foreign country, an FCC-licensed amateur must also transmit the call sign of the foreign station as well as his own call sign.

T1E06 Besides normal identification, what else must a US station do when sending third-party communications internationally?

A. The US station must transmit its own call sign at the beginning of each communication, and at least every ten minutes after that

B. The US station must transmit both call signs at the end of each communication

C. The US station must transmit its own call sign at the beginning of each communication, and at least every five minutes after that

D. Each station must transmit its own call sign at the end of each transmission, and at least every five minutes after that

B If you are sending a message on behalf of someone who is not an amateur to another amateur outside the US, you must **transmit both your call sign and the other station's call sign** at the end of the message exchange. [97.115 (c)]

T1E07 When is an amateur allowed to broadcast information to the general public?

A. Never

B. Only when the operator is being paid

C. Only when broadcasts last less than 1 hour

D. Only when broadcasts last longer than 15 minutes

A Amateur stations are **never** allowed to broadcast information to the general public. [97.113 (b)]

T1E08 When is an amateur station permitted to transmit music?

A. Never, except incidental music during authorized rebroadcasts of space shuttle communications

B. Only if the transmitted music produces no spurious emissions

C. Only if it is used to jam an illegal transmission

D. Only if it is above 1280 MHz, and the music is a live performance

A Amateur stations are never permitted to transmit music. This includes "incidental" background music that may be playing while you are transmitting. The **one exception** is that if an amateur station is **retransmitting audio from the space shuttle** (with NASA permission), incidental background music that may be included with transmissions to or from the shuttle is permitted. [97.113 (a) (4), 97.113 (e)]

T1E09 When is the use of codes or ciphers allowed to hide the meaning of an amateur message?

A. Only during contests
B. Only during nationally declared emergencies
C. Never, except when special requirements are met
D. Only on frequencies above 1280 MHz

C **You may not use codes or ciphers** to send "secret" messages in the Amateur Radio Service. All messages must be in plain language, using only commonly understood abbreviations. The **one exception** is that a **telecommand station** may transmit control messages **to a space station** using special codes or ciphers that are intended to obscure the meaning of the telecommand messages. [97.113 (a) (4), 97.211 (b)]

T1E10 Which of the following one-way communications may not be transmitted in the amateur service?

A. Telecommands to model craft
B. Broadcasts intended for the general public
C. Brief transmissions to make adjustments to the station
D. Morse code practice

B No one is permitted to use frequencies in the Amateur Radio Service to transmit one-way communications intended for the **general public**. This type of transmission is called **broadcasting**. [97.3 (a) (10), 97.113 (b)]

T1E11 If you are allowing a non-amateur friend to use your station to talk to someone in the US, and a foreign station breaks in to talk to your friend, what should you do?

A. Have your friend wait until you find out if the US has a third-party agreement with the foreign station's government
B. Stop all discussions and quickly sign off
C. Since you can talk to any foreign amateurs, your friend may keep talking as long as you are the control operator
D. Report the incident to the foreign amateur's government

A Some governments forbid the exchange of third-party messages, so you can exchange such communications with a foreign amateur only **if the two countries have a third-party agreement**. Since your non-amateur friend is a third party, you must be sure the US has a third-party agreement with the foreign amateur's country before you permit your friend to talk. The list of countries changes from time to time; the ARRL monthly journal, *QST*, publishes an updated list periodically. [97.115 (a) (2)]

T1E12 When are you allowed to transmit a message to a station in a foreign country for a third party?

A. Anytime

B. Never

C. Anytime, unless there is a third-party agreement between the US and the foreign government

D. If there is a third-party agreement with the US government, or if the third party is eligible to be the control operator

D   Some governments forbid the exchange of third-party messages, so you can exchange such communications with a foreign amateur only **if the two countries have a third-party agreement**. If the **third party** is another **amateur who** *could* **be the control operator**, then you can pass the message even if there is no third-party agreement with the other country. [97.115 (a) (2)]

T1F Frequency selection and sharing; transmitter power; digital communications

T1F01 If the FCC rules say that the amateur service is a secondary user of a frequency band, and another service is a primary user, what does this mean?

A. Nothing special; all users of a frequency band have equal rights to operate

B. Amateurs are only allowed to use the frequency band during emergencies

C. Amateurs are allowed to use the frequency band only if they do not cause harmful interference to primary users

D. Amateurs must increase transmitter power to overcome any interference caused by primary users

C   According to the FCC Rules, if a radio service is a **secondary user** of a frequency, stations in that service **must not interfere with** stations in the **primary service**. Likewise, if a radio service is a secondary user, operators must accept any interference from stations in the primary service. [97.303]

T1F02 What rule applies if two amateur stations want to use the same frequency?

A. The station operator with a lesser class of license must yield the frequency to a higher-class licensee

B. The station operator with a lower power output must yield the frequency to the station with a higher power output

C. Both station operators have an equal right to operate on the frequency

D. Station operators in ITU Regions 1 and 3 must yield the frequency to stations in ITU Region 2

C   Frequencies in the amateur bands are assigned to all stations in general. No station is assigned a particular frequency for their own use. All amateur stations have an **equal right to operate** on any frequency within their license privileges. [97.101 (b)]

T1F03 If a repeater is causing harmful interference to another repeater and a frequency coordinator has recommended the operation of one repeater only, who is responsible for resolving the interference?

A. The licensee of the unrecommended repeater

B. Both repeater licensees

C. The licensee of the recommended repeater

D. The frequency coordinator

A   *Frequency coordinators* select and recommend repeater operating frequencies to minimize interference between stations. The FCC Rules do not *require* repeaters to use only coordinated frequencies. If one repeater has a coordinated frequency and the other does not, however, the **licensee** of the **unrecommended repeater** has primary responsibility to resolve interference problems between the two. [97.205 (c)]

T1F04 If a repeater is causing harmful interference to another amateur repeater and a frequency coordinator has recommended the operation of both repeaters, who is responsible for resolving the interference?

    A. The licensee of the repeater that has been recommended for the longest period of time
    B. The licensee of the repeater that has been recommended the most recently
    C. The frequency coordinator
    D. Both repeater licensees

D   Frequency coordinators select and recommend repeater operating frequencies to minimize interference between stations. The FCC Rules do not *require* repeaters to use only coordinated frequencies. If two repeaters use the same frequency, and both have been recommended by the frequency coordinator, **both repeater licensees** share the responsibility to resolve interference problems between them. [97.205 (c)]

T1F05 What is the term for the average power supplied to an antenna transmission line during one RF cycle at the crest of the modulation envelope?

    A. Peak transmitter power
    B. Peak output power
    C. Average radio-frequency power
    D. Peak envelope power

D   FCC Rules set the maximum output power limits in terms of **peak envelope power (PEP)** output from a transceiver. PEP is defined as the *average* power during one RF cycle at the crest (maximum point) of the modulation envelope. [97.3 (b) (6)]

T1F06 What is the maximum transmitting power permitted an amateur station on 146.52 MHz?

    A. 200 watts PEP output
    B. 500 watts ERP
    C. 1000 watts DC input
    D. 1500 watts PEP output

D   Technicians, along with all other licensees, may use a *maximum* **peak envelope power (PEP) output of 1500 watts on the 2-meter band**. We've emphasized the word *maximum* because the FCC Rules also require that "An amateur station must use the minimum power necessary to carry out the desired communications." [97.313 (b)]

T1F07 On which band(s) may a Technician licensee who has passed a Morse code exam use up to 200 watts PEP output power?

A. 80, 40, 15, and 10 meters
B. 80, 40, 20, and 10 meters
C. 1.25 meters
D. 23 centimeters

A Technicians with code credit may use up to 200 watts PEP output power on parts of four different HF bands, **80, 40, 15 and 10 meters**. [97.313 (c)]

T1F08 What amount of transmitter power must amateur stations use at all times?

A. 25 watts PEP output
B. 250 watts PEP output
C. 1500 watts PEP output
D. The minimum legal power necessary to communicate

D Always use the **minimum power necessary** to carry out your desired communications. If you run more power than is really needed you will be more likely to interfere with other stations, and that is a poor operating practice. [97.313 (a)]

T1F09 What name does the FCC use for telemetry, telecommand or computer communications emissions?

A. CW
B. Image
C. Data
D. RTTY

C The FCC defines **data** communications as telemetry, telecommand and computer communications emissions. Packet radio is one example of data communications. [97.3 (c) (2)]

T1F10 What name does the FCC use for narrow-band direct-printing telegraphy emissions?

A. CW
B. Image
C. MCW
D. RTTY

D   The FCC defines **RTTY** (radioteletype) communications as narrow-band direct-printing telegraphy emissions. When stations are communicating by RTTY, one operator is typing the message or information on a keyboard for transmission and other stations receive those signals and display them directly on a computer screen or teleprinter. [97.3 (c) (7)]

T1F11 What is the maximum symbol rate permitted for packet transmissions on the 2-meter band?

A. 300 bauds
B. 1200 bauds
C. 19.6 kilobauds
D. 56 kilobauds

C   Operators transmitting packet radio signals on the 2-meter band may use signaling rates up to **19.6 kilobauds**. The signaling speed in bauds is equal to the reciprocal of the shortest pulse. For a 19.6-kbaud signaling rate the shortest pulse is 1/19600 = 51 microseconds. [97.307 (f) (5)]

T1F12 What is the maximum symbol rate permitted for RTTY or data transmissions on the 6- and 2-meter bands?

A. 56 kilobauds
B. 19.6 kilobauds
C. 1200 bauds
D. 300 bauds

B   Operators transmitting radioteletype (RTTY) or data signals on the amateur 6 and 2-meter bands may use signaling rates up to **19.6 kilobauds**. The signaling speed in bauds is equal to the reciprocal of the shortest pulse. For a 19.6-kbaud signaling rate the shortest pulse is 1/19600 = 51 microseconds. [97.307 (f) (5)]

T1G Satellite and space communications; false signals or unidentified communications; malicious interference

T1G01 What is an amateur space station?

- A. An amateur station operated on an unused frequency
- B. An amateur station awaiting its new call letters from the FCC
- C. An amateur station located more than 50 kilometers above the Earth's surface
- D. An amateur station that communicates with the International Space Station

C   An Amateur Radio space station is **any station that is more than 50 km** (31 miles) **above the Earth.** Any satellite, including the space shuttle and the International Space Station, is an amateur space station if it operates on Amateur Radio frequencies. [97.3 (a) (40)]

T1G02 Who may be the licensee of an amateur space station?

- A. An amateur holding an Amateur Extra class operator license
- B. Any licensed amateur operator
- C. Anyone designated by the commander of the spacecraft
- D. No one unless specifically authorized by the government

B   All it takes to be the licensee of an amateur space station is your operator license. **Any licensed amateur** may be the licensee of a space station. So if you become an astronaut, and have the opportunity to fly aboard the space shuttle or International Space Station, plan to take some Amateur Radio equipment along! [97.207 (a)]

T1G03 Which band may NOT be used by Earth stations for satellite communications?

- A. 6 meters
- B. 2 meters
- C. 70 centimeters
- D. 23 centimeters

A   *Earth stations* are stations located on the Earth's surface, or within 50 km of it. They are intended for communications with *space stations*, or with other Earth stations by means of one or more objects (satellites) in space. Any amateur can be the control operator of an Earth station. Some specific transmitting frequencies are authorized for Earth stations. These are located in the 40, 20, 17, 15, 12, 10 and 2-meter bands, in the 23 and 70-cm bands and in other amateur bands that are higher in frequency. There are **no frequencies** available for Earth stations **in the 6-meter band**, however. [97.209 (b)]

T1G04 When may false or deceptive amateur signals or communications be transmitted?

    A. Never
    B. When operating a beacon transmitter in a "fox hunt" exercise
    C. When playing a harmless "practical joke"
    D. When you need to hide the meaning of a message for secrecy

A   You should **never** transmit any kind of message that is intended to deceive someone who may be listening. This means you should never transmit an emergency call for help if you are not in need of such assistance. You should never transmit another station's call sign in a way that will make others think you are that station. [97.113 (a) (4)]

T1G05 If an amateur pretends there is an emergency and transmits the word "MAYDAY," what is this called?

    A. A traditional greeting in May
    B. An emergency test transmission
    C. False or deceptive signals
    D. Nothing special; "MAYDAY" has no meaning in an emergency

C   MAYDAY is an international distress signal, and means there is a life-threatening emergency. If an amateur transmits the word MAYDAY when no emergency exists, it is a **false or deceptive signal**, and that is illegal. [97.113 (a) (4)]

T1G06 When may an amateur transmit unidentified communications?

    A. Only for brief tests not meant as messages
    B. Only if it does not interfere with others
    C. Never, except transmissions from a space station or to control a model craft
    D. Only for two-way or third-party communications

C   You must always identify your amateur station. You may use amateur frequencies to **transmit control signals to a model craft**, however, with a transmitter power of less than 1 watt, and in that case the station identification rules do not apply. In addition, **space stations are not required to transmit identification**. [97.119 (a)]

T1G07 What is an amateur communication called that does not have the required station identification?

A. Unidentified communications or signals
B. Reluctance modulation
C. Test emission
D. Tactical communication

A   If an amateur operator fails to transmit the proper station identification, the transmission is an **unidentified communication** or an **unidentified signal**. [97.119 (a)]

T1G08 If an amateur transmits to test access to a repeater without giving any station identification, what type of communication is this called?

A. A test emission; no identification is required
B. An illegal unmodulated transmission
C. An illegal unidentified transmission
D. A non-communication; no voice is transmitted

C   You must always identify your amateur station. If an amateur operator fails to transmit the proper station identification, the transmission is an **illegal unidentified communication** or an **unidentified signal**. [97.119 (a)]

T1G09 When may you deliberately interfere with another station's communications?

A. Only if the station is operating illegally
B. Only if the station begins transmitting on a frequency you are using
C. Never
D. You may expect, and cause, deliberate interference because it can't be helped during crowded band conditions

C   You should **never** deliberately interfere with another station. If you are aware that your transmissions are interfering with another station you should take steps to correct the problem. [97.101 (d)]

T1G10 If an amateur repeatedly transmits on a frequency already occupied by a group of amateurs in a net operation, what type of interference is this called?

A. Break-in interference
B. Harmful or malicious interference
C. Incidental interference
D. Intermittent interference

B   Amateurs may not deliberately interfere with other amateur communication. An amateur who repeatedly transmits on a frequency already occupied by a group of amateurs is causing **harmful or malicious interference**. [97.3(a) (22)]

T1G11 What is a transmission called that disturbs other communications?

A. Interrupted CW
B. Harmful interference
C. Transponder signals
D. Unidentified transmissions

B   Any transmitted signal that disturbs, or interferes with, another communication is called **harmful interference**. [97.3 (a) (23)]

T1H   Correct language; phonetics; beacons; radio ontrol of model craft and vehicles

T1H01 If you are using a language besides English to make a contact, what language must you use when identifying your station?

A. The language being used for the contact
B. The language being used for the contact, provided the US has a third-party communications agreement with that country
C. English
D. Any language of a country that is a member of the International Telecommunication Union

C   The FCC Rules require you to identify your station in **English**. You may use any language to talk with another station, as long as your station ID is given in English. [97.119 (b) (2)]

T1H02 What do the FCC Rules suggest you use as an aid for correct station identification when using phone?

A. A speech compressor
B. Q signals
C. A phonetic alphabet
D. Unique words of your choice

C   The FCC Rules suggest that you use a standard **phonetic alphabet** when you identify your station using voice (phone). Many amateurs like to use funny phrases or words as phonetics for their call sign. There is nothing illegal about this, but the standard International Telecommunication Union (ITU) phonetic alphabet was carefully designed to be easily understood by someone speaking nearly any language around the world. [97.119 (b) (2)]

T1H03 What is the advantage in using the International Telecommunication Union (ITU) phonetic alphabet when identifying your station?

A. The words are internationally recognized substitutes for letters
B. There is no advantage
C. The words have been chosen to be easily pronounced by Asian cultures
D. It preserves traditions begun in the early days of Amateur Radio

A   The FCC Rules suggest that you use a standard phonetic alphabet when you identify your station using voice (phone). Many amateurs like to use funny phrases or words as phonetics for their call sign. There is nothing illegal about this, but the standard International Telecommunication Union (ITU) phonetic alphabet was carefully designed to be easily understood by someone speaking nearly any language around the world. These words are **internationally recognized substitutes for letters**. [97.119 (b) (2)]

Standard ITU Phonetics

A—Alfa (**AL** FAH)
B—Bravo (**BRAH** VOH)
C—Charlie (**CHAR** LEE) or (**SHAR** LEE)
D—Delta (**DELL** TAH)
E—Echo (**ECK** OH)
F—Foxtrot (**FOKS** TROT)
G—Golf (GOLF)
H—Hotel (HOH **TELL**)
I—India (**IN** DEE AH)
J—Juliett (**JEW** LEE ETT)
K—Kilo (**KEY** LOH)
L—Lima (**LEE** MAH)
M—Mike (MIKE)
N—November (NO **VEM** BER)
O—Oscar (**OSS** CAH)
P—Papa (PAH **PAH**)
Q—Quebec (KEH **BECK**)
R—Romeo (**ROW** ME OH)
S—Sierra (SEE **AIR** RAH)
T—Tango (**TANG** GO)
U—Uniform (**YOU** NEE FORM) or (OO NEE FORM)
V—Victor (**VIK** TAH)
W—Whiskey (**WISS** KEY)
X—X-RAY (**ECKS** RAY)
Y—Yankee (**YANG** KEY)
Z—Zulu (**ZOO** LOO)

Note: The **boldfaced** syllables are emphasized. The pronunciations shown in this table were designed for those who speak any of the international languages. The pronunciations given for "Oscar" and "Victor" may seem awkward to English-speaking people in the US.

T1H04 What is one reason to avoid using "cute" phrases or word combinations to identify your station?

A. They are not easily understood by non-English-speaking amateurs

B. They might offend English-speaking amateurs

C. They do not meet FCC identification requirements

D. They might be interpreted as codes or ciphers intended to obscure the meaning of your identification

A The FCC Rules suggest that you use a standard phonetic alphabet when you identify your station using voice (phone). Many amateurs like to use funny phrases or words as phonetics for their call sign. There is nothing illegal about this, but many **amateurs who do not speak English may not understand the words** or the humor. The standard International Telecommunication Union (ITU) phonetic alphabet was carefully designed to be easily understood by someone speaking nearly any language around the world. These words are internationally recognized substitutes for letters. [97.119 (b) (2)]

T1H05 What is an amateur station called that transmits communications for the purpose of observation of propagation and reception?

A. A beacon

B. A repeater

C. An auxiliary station

D. A radio control station

A **Beacon stations** transmit communications to help amateurs observe propagation conditions and the reception of signals from various locations around the world. Try listening to the world-wide beacon system on 14.1 MHz. Stations transmit their call signs and then a long dash at powers of 100, 10, 1 and 0.1 watt. [97.3 (a) (9)]

T1H06 What is the maximum transmitting power permitted an amateur station in beacon operation?

A. 10 watts PEP output

B. 100 watts PEP output

C. 500 watts PEP output

D. 1500 watts PEP output

B Stations in beacon operation are permitted a maximum output power of **100 watts PEP**. [97.203 (c)]

T1H07 What minimum class of amateur license must you hold to operate a beacon or a repeater station?

    A. Technician with credit for passing a Morse code exam
    B. Technician
    C. General
    D. Amateur Extra

B   Any licensed amateur with a **Technician** or higher license may be the control operator of a beacon station or a repeater station. [97.203 (a), 97.205 (a)]

T1H08 What minimum information must be on a label affixed to a transmitter used for telecommand (control) of model craft?

    A. Station call sign
    B. Station call sign and the station licensee's name
    C. Station call sign and the station licensee's name and address
    D. Station call sign and the station licensee's class of license

C   You may use Amateur Radio frequencies to transmit telecommand (control) signals for remote control of model craft if the transmitter output power is less than 1 watt. Such a station does not have to transmit the station call sign, but there must be a label attached to the transmitter that lists the **station call sign** and the **licensee's name and address**. [97.215 (a)]

T1H09 What is the maximum transmitter power an amateur station is allowed when used for telecommand (control) of model craft?

    A. One milliwatt
    B. One watt
    C. 25 watts
    D. 100 watts

B   You may use Amateur Radio frequencies to transmit telecommand (control) signals for remote control of model craft if the transmitter output power is less than **one watt**. [97.215 (c)]

T1I Emergency communications; broadcasting; indecent and obscene language

T1I01 If you hear a voice distress signal on a frequency outside of your license privileges, what are you allowed to do to help the station in distress?

A. You are NOT allowed to help because the frequency of the signal is outside your privileges

B. You are allowed to help only if you keep your signals within the nearest frequency band of your privileges

C. You are allowed to help on a frequency outside your privileges only if you use international Morse code

D. You are allowed to help on a frequency outside your privileges in any way possible

D During an emergency involving immediate protection of life or property, **you can do just about anything to carry out the necessary communications!** The FCC Rules state: "No provision of these rules prevents the use by an amateur station in distress of any means at its disposal to attract attention, make known its condition and location, and obtain assistance. " All of this means **you can use a frequency outside of your normal privileges** and use a mode you might not normally be allowed to use if it will help complete the emergency communications. [97.405 (a)]

T1I02 When may you use your amateur station to transmit an "SOS" or "MAYDAY"?

A. Never

B. Only at specific times (at 15 and 30 minutes after the hour)

C. In a life- or property-threatening emergency

D. When the National Weather Service has announced a severe weather watch

C The Morse code international distress call is SOS and the distress call for voice modes is MAYDAY. You should **only use these calls** to attract attention **during** a real emergency that involves a **life or property-threatening emergency**. [97.403]

T1I03 When may you send a distress signal on any frequency?

    A. Never

    B. In a life- or property-threatening emergency

    C. Only at specific times (at 15 and 30 minutes after the hour)

    D. When the National Weather Service has announced a severe weather watch

B   The Morse code international distress call is SOS and the distress call for voice modes is MAYDAY. You should **only use these calls** to attract attention **during** a real emergency that involves a **life or property-threatening emergency**. [97.405 (a)]

T1I04 If a disaster disrupts normal communication systems in an area where the amateur service is regulated by the FCC, what kinds of transmissions may stations make?

    A. Those that are necessary to meet essential communication needs and facilitate relief actions

    B. Those that allow a commercial business to continue to operate in the affected area

    C. Those for which material compensation has been paid to the amateur operator for delivery into the affected area

    D. Those that are to be used for program production or news gathering for broadcasting purposes

A   Amateur Radio operators have a long history of providing emergency communications, especially when a disaster disrupts normal communications lines. During such a communications emergency, amateurs must limit their operation in the disaster area to those **communications necessary** to **meet essential communication needs** and **facilitate relief actions**. [97.401 (a)]

T1I05 What information is included in an FCC declaration of a temporary state of communication emergency?

    A. A list of organizations authorized to use radio communications in the affected area

    B. A list of amateur frequency bands to be used in the affected area

    C. Any special conditions and special rules to be observed during the emergency

    D. An operating schedule for authorized amateur emergency stations

C   When a disaster disrupts normal communication systems in a particular area, the FCC may declare a temporary state of communication emergency. Such a declaration gives any **special conditions** and **special rules to be observed** by amateurs **during the** communication **emergency**. [97.401 (c)]

T1106 What is meant by the term broadcasting?

A. Transmissions intended for reception by the general public, either direct or relayed
B. Retransmission by automatic means of programs or signals from non-amateur stations
C. One-way radio communications, regardless of purpose or content
D. One-way or two-way radio communications between two or more stations

A   Broadcasting involves sending a **transmission to the general public**. Amateurs transmit messages to a particular ham radio operator or group of ham radio operators. [97.3 (a) (10)]

T1107 When may you send obscene words from your amateur station?

A. Only when they do not cause interference to other communications
B. Never; obscene words are not allowed in amateur transmissions
C. Only when they are not retransmitted through a repeater
D. Any time, but there is an unwritten rule among amateurs that they should not be used on the air

B   Signals from your Amateur Radio station can travel around the world, and be picked up by listeners of any age. You should **never transmit obscene words** because it may offend someone who is listening and because it is against FCC Rules. There is no list of forbidden words, so your own judgment is your best guide. If there is even a remote possibility that someone might be offended by your language, don't say it! [97.113 (a) (4)]

T1108 When may you send indecent words from your amateur station?

A. Only when they do not cause interference to other communications
B. Only when they are not retransmitted through a repeater
C. Any time, but there is an unwritten rule among amateurs that they should not be used on the air
D. Never; indecent words are not allowed in amateur transmissions

D   Signals from your Amateur Radio station can travel around the world, and be picked up by listeners of any age. You should **never transmit indecent words** because it may offend someone who is listening and because it is against FCC Rules. There is no list of forbidden words, so your own judgment is your best guide. If there is even a remote possibility that someone might be offended by your language, don't say it! [97.113 (a) (4)]

T1I09 Why is indecent and obscene language prohibited in the Amateur Service?

    A. Because it is offensive to some individuals

    B. Because young children may intercept amateur communications with readily available receiving equipment

    C. Because such language is specifically prohibited by FCC Rules

    D. All of these choices are correct

    D   Amateurs should never transmit any language that could be interpreted to be indecent or obscene. Aside from being **offensive** to adults, such language could be overheard by **young children**. In addition, **FCC Rules** specifically **prohibit indecent or obscene language**. [97.113 (a) (4)]

T1I10 Where can the official list of prohibited obscene and indecent words be found?

    A. There is no public list of prohibited obscene and indecent words; if you believe a word is questionable, don't use it in your communications

    B. The list is maintained by the Department of Commerce

    C. The list is International, and is maintained by Industry Canada

    D. The list is in the "public domain," and can be found in all amateur study guides

    A   Although there is **no official list of obscene or indecent words**, if you are unsure of whether a word falls under those categories, *don't use it on the air!* [97.113 (a) (4)]

T1I11 Under what conditions may a Technician class operator use his or her station to broadcast information intended for reception by the general public?

    A. Never, broadcasting is a privilege reserved for Extra and General class operators only

    B. Only when operating in the FM Broadcast band (88.1 to 107.9 MHz)

    C. Only when operating in the AM Broadcast band (530 to 1700 kHz)

    D. Never, broadcasts intended for reception by the general public are not permitted in the Amateur Service

    D   Amateur stations are **never** allowed to broadcast information to the general public. [97.113 (b)]

T2

Operating Procedures

There will be 5 questions on your Technician exam from the Operating Procedures subelement. Those 5 questions will be taken from the 5 groups of questions, labeled T2A through T2E, printed in this chapter.

T2A Preparing to transmit; choosing a frequency for tune-up; operating or emergencies; morse code; repeater operations and autopatch

T2A01 What should you do before you transmit on any frequency?

A. Listen to make sure others are not using the frequency
B. Listen to make sure that someone will be able to hear you
C. Check your antenna for resonance at the selected frequency
D. Make sure the SWR on your antenna feed line is high enough

A **Always listen before you transmit. Make sure someone** else **isn't** already **using the frequency** on which you want to operate. On the HF bands, it is courteous to ask, "Is the frequency in use?" or to send the Morse code signal "QRL?"

T2A02 If you are in contact with another station and you hear an emergency call for help on your frequency, what should you do?

A. Tell the calling station that the frequency is in use
B. Direct the calling station to the nearest emergency net frequency
C. Call your local Civil Preparedness Office and inform them of the emergency
D. Stop your QSO immediately and take the emergency call

D Emergency communications have the highest priority. If you hear an emergency call for help **stop whatever you are doing and take down the message.** Try to acknowledge the transmitting station to let them know that you copied the message, and then contact the appropriate authorities.

T2A03 Why should local amateur communications use VHF and UHF frequencies instead of HF frequencies?

    A. To minimize interference on HF bands capable of long-distance communication
    B. Because greater output power is permitted on VHF and UHF
    C. Because HF transmissions are not propagated locally
    D. Because signals are louder on VHF and UHF frequencies

A   If you want to communicate with other stations across the country or around the world, you should use the high frequency (HF) bands. For local communications it is better to use very-high frequencies (VHF) or ultra-high frequencies (UHF), to **minimize the interference** levels **on the HF bands**.

T2A04 How can on-the-air interference be minimized during a lengthy transmitter testing or loading-up procedure?

    A. Choose an unoccupied frequency
    B. Use a dummy load
    C. Use a non-resonant antenna
    D. Use a resonant antenna that requires no loading-up procedure

B   You should always use a **dummy antenna**, or **dummy load**, when you are testing or tuning your transmitter. This will minimize the amount of on-the-air interference you will create.

T2A05 At what speed should a Morse code CQ call be transmitted?

    A. Only speeds below five WPM
    B. The highest speed your keyer will operate
    C. Any speed at which you can reliably receive
    D. The highest speed at which you can control the keyer

C   Most people can transmit Morse code faster than they can reliably receive it. It is a good operating procedure to **send a CQ call at a speed you can receive**, and to reply to a CQ call at a speed no faster than the calling station used.

T2A06 What is an autopatch?

A. An automatic digital connection between a US and a foreign amateur
B. A digital connection used to transfer data between a hand-held radio and a computer
C. A device that allows radio users to access the public telephone system
D. A video interface allowing images to be patched into a digital data stream

C   An autopatch provides a connection between a repeater and a telephone line. This allows hams to **access the public telephone system** so they can make personal telephone calls through a repeater.

T2A07 How do you call another station on a repeater if you know the station's call sign?

A. Say "break, break 79," then say the station's call sign
B. Say the station's call sign, then identify your own station
C. Say "CQ" three times, then say the station's call sign
D. Wait for the station to call "CQ," then answer it

B   Calling another station on a repeater is a simple procedure. Just **give the other station's call sign**, and **then identify your own station**. "N1PUR this is WR1B."

T2A08 What is a courtesy tone (used in repeater operations)?

A. A sound used to identify the repeater
B. A sound used to indicate when a transmission is complete
C. A sound used to indicate that a message is waiting for someone
D. A sound used to activate a receiver in case of severe weather

B   Many repeaters use a courtesy tone, or **beep to indicate when a station stops transmitting** to the repeater. This allows listening stations an opportunity to join the conversation by sending their call signs.

T2A09 What is the meaning of the procedural signal "DE"?

A. "From" or "this is," as in "WØAIH DE KA9FOX"
B. "Directional Emissions" from your antenna
C. "Received all correctly"
D. "Calling any station"

A   The procedural signal DE is an abbreviated way to say **"from"** or **"this is"** when using Morse code (CW) or radioteletype (RTTY).

T2A10 During commuting rush hours, which type of repeater operation should be discouraged?

A. Mobile stations
B. Low-power stations
C. Highway traffic information nets
D. Third-party communications nets

D   Commuting rush hours are busy times for most repeaters. Many mobile stations are talking on, or listening to, the repeaters. This is generally *not* a good time to operate **third-party communications nets**.

T2A11 What is the proper way to break into a conversation on a repeater?

A. Wait for the end of a transmission and start calling the desired party
B. Shout, "break, break!" to show that you're eager to join the conversation
C. Turn on an amplifier and override whoever is talking
D. Say your call sign during a break between transmissions

D   If you want to join a conversation on a repeater simply wait for the stations to pause between transmissions, and then **say your call sign**. Avoid using the common Citizen's Band (CB) slang "break break" or "breaker" when you want to join the conversation. On most repeaters, the word break is reserved for emergency use.

T2B   Definition and proper use; courteous operation; repeater frequency coordination; morse code

T2B01 When using a repeater to communicate, which of the following do you need to know about the repeater?

A. Its input frequency and offset
B. Its call sign
C. Its power level
D. Whether or not it has an autopatch

A   Repeaters receive on one frequency (the **input frequency**) and transmit on another frequency (the output frequency). The input and output frequencies are separated by a certain amount that is different for each band. This separation is called the **offset**. To use a repeater, your station must transmit on the repeater input frequency and receive on the repeater output frequency. Repeater contacts are often called *duplex operation*, because two frequencies are used. Most radios designed for use on repeaters require you to set the repeater *output* frequency (which is the frequency you listen on) and the offset. The radio then automatically switches to the correct *input* frequency when you transmit.

T2B02 What is an autopatch?

A. Something that automatically selects the strongest signal to be repeated
B. A device that connects a mobile station to the next repeater if it moves out of range of the first
C. A device that allows repeater users to make telephone calls from their stations
D. A device that locks other stations out of a repeater when there is an important conversation in progress

C An autopatch provides a connection between a repeater and a telephone line, and **allows repeater users to make personal telephone calls** through a repeater.

T2B03 What is the purpose of a repeater time-out timer?

A. It lets a repeater have a rest period after heavy use
B. It logs repeater transmit time to predict when a repeater will fail
C. It tells how long someone has been using a repeater
D. It limits the amount of time someone can transmit on a repeater

D A time-out timer limits the length of time a repeater will transmit, and is a safeguard against having the repeater accidentally transmit continuously. A time-out timer also **limits the amount of time you can transmit without pausing** to allow other operators to use the repeater.

T2B04 What is a CTCSS (or PL) tone?

A. A special signal used for telecommand control of model craft
B. A sub-audible tone, added to a carrier, which may cause a receiver to accept a signal
C. A tone used by repeaters to mark the end of a transmission
D. A special signal used for telemetry between amateur space stations and Earth stations

B A continuous-tone-coded squelch system (CTCSS) on a receiver requires a particular **low-frequency audio tone to open the receiver squelch**. Unless the transmitted signal includes the proper tone, the receiver will ignore the signal. Since the tones are less than 300 Hz you won't be able to hear them on the signal, so they are called sub-audible. Private Line (PL) is a Motorola trademark for CTCSS.

T2B05 What is the usual input/output frequency separation for repeaters in the 2-meter band?

A. 600 kHz
B. 1.0 MHz
C. 1.6 MHz
D. 5.0 MHz

A   Repeaters use a pair of frequencies, called the *input* and the *output* frequencies. Stations transmit to the repeater on the input frequency and listen to the signal from the repeater on the output frequency. On the 2-meter band (144 to 148 MHz), most repeaters have a **600-kHz** separation between their input and output frequencies.

T2B06 What is the usual input/output frequency separation for repeaters in the 1.25-meter band?

A. 600 kHz
B. 1.0 MHz
C. 1.6 MHz
D. 5.0 MHz

C   Repeaters use a pair of frequencies, called the *input* and the *output* frequencies. Stations transmit to the repeater on the input frequency and listen to the signal from the repeater on the output frequency. On the 1.25-meter band (222 to 225 MHz), most repeaters have a **1.6-MHz** separation between their input and output frequencies.

T2B07 What is the usual input/output frequency separation for repeaters in the 70-centimeter band?

A. 600 kHz
B. 1.0 MHz
C. 1.6 MHz
D. 5.0 MHz

D   Repeaters use a pair of frequencies, called the *input* and the *output* frequencies. Stations transmit to the repeater on the input frequency and listen to the signal from the repeater on the output frequency. On the 70-centimeter band (420 to 450 MHz), most repeaters have a **5-MHz** separation between their input and output frequencies.

T2B08 What is the purpose of repeater operation?

A. To cut your power bill by using someone else's higher power system
B. To help mobile and low-power stations extend their usable range
C. To transmit signals for observing propagation and reception
D. To communicate with stations in services other than amateur

B   Repeaters receive signals from stations, usually operating on VHF or UHF, and retransmit the signals with higher power from a location that **extends their communications range**. Repeaters are especially useful for **low-power portable and mobile stations**.

T2B09 What is a repeater called that is available for anyone to use?

A. An open repeater
B. A closed repeater
C. An autopatch repeater
D. A private repeater

A   An **open repeater** is one that is available for all operators to use, whether or not the operator is a member of the local club or repeater organization.

T2B10 Why should you pause briefly between transmissions when using a repeater?

A. To check the SWR of the repeater
B. To reach for pencil and paper for third-party communications
C. To listen for anyone wanting to break in
D. To dial up the repeater's autopatch

C   When you are using a repeater, it is a good practice to pause briefly between transmissions to listen for any other stations that may want to join your conversation or call another station.

T2B11 Why should you keep transmissions short when using a repeater?

A. A long transmission may prevent someone with an emergency from using the repeater
B. To see if the receiving station operator is still awake
C. To give any listening non-hams a chance to respond
D. To keep long-distance charges down

A When you are communicating with other stations over a repeater you should keep your transmissions short. Many amateurs depend on repeaters for emergency communications, especially while they are mobile, so long transmissions may delay or prevent an emergency call for help.

T2C Simplex operations; RST signal reporting; choice of equipment for desired communications; communications modes including amateur television (ATV), packet radio; Q signals, procedural signals and abbreviations

T2C01 What is simplex operation?

A. Transmitting and receiving on the same frequency
B. Transmitting and receiving over a wide area
C. Transmitting on one frequency and receiving on another
D. Transmitting one-way communications

A Simplex operation means the stations are **talking to each other directly, on one frequency**. The term usually refers to contacts on the VHF and UHF bands without the use of a repeater.

T2C02 When should you use simplex operation instead of a repeater?

A. When the most reliable communications are needed
B. When a contact is possible without using a repeater
C. When an emergency telephone call is needed
D. When you are traveling and need some local information

B If you are talking with a nearby amateur, and **communication is possible without using a repeater**, switch to simplex operation.

T2C03 Why should simplex be used where possible, instead of using a repeater?

A. Signal range will be increased
B. Long distance toll charges will be avoided
C. The repeater will not be tied up unnecessarily
D. Your antenna's effectiveness will be better tested

C   Considerate operators use a *simplex* frequency (both stations transmit and receive on the same frequency) whenever possible for VHF or UHF communications. This leaves the repeater frequencies clear for possible emergency calls, or for stations that are not able to use a simplex frequency. **Don't tie up the repeater unnecessarily.**

T2C04 If you are talking to a station using a repeater, how would you find out if you could communicate using simplex instead?

A. See if you can clearly receive the station on the repeater's input frequency
B. See if you can clearly receive the station on a lower frequency band
C. See if you can clearly receive a more distant repeater
D. See if a third station can clearly receive both of you

A   Many VHF and UHF FM transceivers have a REVERSE button, which allows you to listen on the repeater input frequency. Press this button while the other station is transmitting to determine if you can hear them "direct," without the repeater. If you can **clearly receive** the other station on the **repeater input frequency**, then you can probably continue your communication on simplex.

T2C05 What does RST mean in a signal report?

A. Recovery, signal strength, tempo
B. Recovery, signal speed, tone
C. Readability, signal speed, tempo
D. Readability, signal strength, tone

D   The **Readability** of a signal is expressed on a scale of 1 to 5, signal **Strength** is given on a scale from 1 to 9, usually read from the receiver S(ignal strength) meter and **Tone** describes the signal tone quality on a scale from 1 to 9. Tone is only used with Morse code (CW), radioteletype (RTTY) and data signals.

The RST System

READABILITY
1—Unreadable.
2—Barely readable, occasional words distinguishable.
3—Readable with considerable difficulty.
4—Readable with practically no difficulty.
5—Perfectly readable.

SIGNAL STRENGTH
1—Faint signals barely perceptible.
2—Very weak signals.
3—Weak signals.
4—Fair signals.
5—Fairly good signals.
6—Good signals.
7—Moderately strong signals.
8—Strong signals.
9—Extremely strong signals.

TONE
1—Sixty-cycle ac or less, very rough and broad.
2—Very rough ac, very harsh and broad.
3—Rough ac tone, rectified but not filtered.
4—Rough note, some trace of filtering.
5—Filtered rectified ac but strongly ripple-modulated.
6—Filtered tone, definite trace of ripple modulation.
7—Near pure tone, trace of ripple modulation.
8—Near perfect tone, slight trace of modulation.
9—Perfect tone, no trace of ripple or modulation of any kind.

The "tone" report refers only to the purity of the signal. It has no connection with its stability or freedom from clicks or chirps. Most of the signals you hear will be a T-9. Other tone reports occur mainly if the power supply filter capacitors are not doing a thorough job. If so, some trace of ac ripple finds its way onto the transmitted signal. If the signal has the characteristic steadiness of crystal control, add X to the report (for example, RST 469X). If it has a chirp or "tail" (either on "make" or "break") add C (for example, 469C). If it has clicks or noticeable other keying transients, add K (for example, 469K). Of course a signal could have both chirps and clicks, in which case both C and K could be used (for example, RST 469CK).

T2C06 What is the meaning of: "Your signal report is five nine plus 20 dB..."?

A. Your signal strength has increased by a factor of 100
B. Repeat your transmission on a frequency 20 kHz higher
C. The bandwidth of your signal is 20 decibels above linearity
D. A relative signal-strength meter reading is 20 decibels greater than strength 9

D   See the explanation for question T2C05. Amateur Radio operators use the RST signal reporting system to describe the *Readability*, *Strength* and *Tone* of a Morse code (CW) signal. On voice modes, only the R and S portions of the report are used. A signal report of "five nine plus 20 dB" means the signal is perfectly readable and has a strength of **20 decibels (dB) greater than an S9 signal** on the receiver's S meter.

T2C07 What is the meaning of the procedural signal "CQ"?

A. "Call on the quarter hour"
B. "New antenna is being tested" (no station should answer)
C. "Only the called station should transmit"
D. "Calling any station"

D   We use the signal "CQ" to establish communications with any operator who may be listening. CQ means "**calling any station**."

T2C08 What is a QSL card in the amateur service?

A. A letter or postcard from an amateur pen pal
B. A Notice of Violation from the FCC
C. A written acknowledgment of communications between two amateurs
D. A postcard reminding you when your license will expire

C   The international Q signal QSL means "**to confirm**," and a QSL card is a post card or other **written proof that a radio contact actually took place**. It is fun to collect QSL cards from the stations you contact on the air. QSL cards are used to prove that you are eligible for many awards, such as the ARRL Worked All States award.

QSLs for K2BSA/1 to Larry Wolfgang, WR1B, 30 Cottage Road, Bozrah, CT 06334.
Operators were from the Radio Amateur Society of Norwich and the Tri-City Amateur Radio Club. Thanks to Dan Dansby, W5URI, Trustee of K2BSA.

T2C09 What is the correct way to call CQ when using voice?

    A. Say "CQ" once, followed by "this is," followed by your call sign spoken three times

    B. Say "CQ" at least five times, followed by "this is," followed by your call sign spoken once

    C. Say "CQ" three times, followed by "this is," followed by your call sign spoken three times

    D. Say "CQ" at least ten times, followed by "this is," followed by your call sign spoken once

C   When you call CQ on voice, follow the **"three-by-three rule."** Say **CQ three times**, then **"this is"** followed by **your call sign three times**. You can end the call by saying "Listening" or something similar so other operators who heard you will know you are about to stop transmitting and listen for their call. You might want to say your call sign once using the international phonetic alphabet. We don't normally call CQ on VHF and UHF repeaters. Just give your call and say "listening" for repeater operation.

T2C10 How should you answer a voice CQ call?

    A. Say the other station's call sign at least ten times, followed by "this is," then your call sign at least twice

    B. Say the other station's call sign at least five times phonetically, followed by "this is," then your call sign at least once

    C. Say the other station's call sign at least three times, followed by "this is," then your call sign at least five times phonetically

    D. Say the other station's call sign once, followed by "this is," then your call sign given phonetically

D   When you answer a voice CQ call you only need to **give the other station's call sign once** and **your call sign one or two times. Give your call sign using the international phonetic alphabet**.

T2C11 What is the meaning of: "Your signal is full quieting..."?

    A. Your signal is strong enough to overcome all receiver noise

    B. Your signal has no spurious sounds

    C. Your signal is not strong enough to be received

    D. Your signal is being received, but no audio is being heard

A   Amateurs don't normally use the RST signal reporting system to describe signals received over a repeater. Instead, we describe the signal in terms of the amount of receiver noise that is heard with the signal. A **signal** that is "full quieting" is **strong enough to overcome all receiver noise**.

T2D Distress calling and emergency drills and communications — operations and equipment; Radio Amateur Civil Emergency Service (RACES)

T2D01 What is the proper distress call to use when operating phone?

- A. Say "MAYDAY" several times
- B. Say "HELP" several times
- C. Say "EMERGENCY" several times
- D. Say "SOS" several times

A   If you are using a phone (voice) transmitter to call for help in a life or property-threatening emergency, use the international voice distress signal **MAYDAY**. Also be sure to give your call sign and details about your emergency, such as the nature of the problem, and your location.

T2D02 What is the proper distress call to use when operating CW?

- A. MAYDAY
- B. QRRR
- C. QRZ
- D. SOS

D   If you are using a Morse code (CW) transmitter to call for help in a life or property-threatening emergency, use the international CW distress signal **SOS**. This is sent as a single Morse code character, without spaces between the letters. Also send your call sign and some details of your emergency.

T2D03 What is the proper way to interrupt a repeater conversation to signal a distress call?

- A. Say "BREAK" twice, then your call sign
- B. Say "HELP" as many times as it takes to get someone to answer
- C. Say "SOS," then your call sign
- D. Say "EMERGENCY" three times

A   If you are making an emergency, or distress, call on a repeater and there is a conversation on the repeater you should say "**Break Break**" and then give your call sign during a pause between the operators. If you are using a repeater and hear a station say "**Break Break**" you should immediately acknowledge the station and stand by for details about their emergency.

T2D04 What is one reason for using tactical call signs such as "command post" or "weather center" during an emergency?

    A. They keep the general public informed about what is going on
    B. They are more efficient and help coordinate public-service communications
    C. They are required by the FCC
    D. They increase goodwill between amateurs

B   *Tactical call signs* such as "command post" or "weather center" are often used during emergency and public service communications because the actual operators at the locations may change, but the function remains the same. Use of such tactical call signs produces **more efficient communication** and makes it easier for the public-service personnel to understand. Of course you must still identify your station using your amateur call sign as required by FCC Rules.

T2D05 What type of messages concerning a person's well-being are sent into or out of a disaster area?

    A. Routine traffic
    B. Tactical traffic
    C. Formal message traffic
    D. Health and Welfare traffic

D   Communications into and out of a disaster area often involve messages about the health and well being of people in the area. Such communications are called **Health and Welfare traffic**.

T2D06 What are messages called that are sent into or out of a disaster area concerning the immediate safety of human life?

    A. Tactical traffic
    B. Emergency traffic
    C. Formal message traffic
    D. Health and Welfare traffic

B   Messages sent into or out of a disaster area that have a direct bearing on the protection of life or property have the highest priority. Such communications are called **Emergency traffic**. Other types of messages, such as Health and Welfare inquiries must wait until the emergency traffic has been handled.

T2D07 Why is it a good idea to have a way to operate your amateur station without using commercial AC power lines?

    A. So you may use your station while mobile
    B. So you may provide communications in an emergency
    C. So you may operate in contests where AC power is not allowed
    D. So you will comply with the FCC rules

B   It is a good idea to have some way to operate your amateur station without the use of commercial ac power because most natural and man-made disasters will knock down power lines. The only way for you to **provide emergency communications** in such a situation is to have an alternative means of powering your station.

T2D08 What is the most important accessory to have for a hand-held radio in an emergency?

    A. An extra antenna
    B. A portable amplifier
    C. Several sets of charged batteries
    D. A microphone headset for hands-free operation

C   Hand-held radios are very helpful for providing emergency communications because you can easily carry the radio into the disaster area without the need for additional equipment. Since you may be without commercial ac power, however, it is important to have **several sets of fully charged batteries** ready to take along with you.

T2D09 Which type of antenna would be a good choice as part of a portable HF amateur station that could be set up in case of an emergency?

    A. A three-element quad
    B. A three-element Yagi
    C. A dipole
    D. A parabolic dish

C   Many amateurs put together an emergency kit, which they can pick up and take to a disaster location on short notice. Such a kit may include equipment for VHF/UHF operation for local communication as well as an HF station for longer-distance communication. A **dipole antenna** is a very good type of HF antenna to include in such a kit because it is easy to set up and use.

T2D10 What is the maximum number of hours allowed per week for RACES drills?

A. One
B. Seven, but not more than one hour per day
C. Eight
D. As many hours as you want

A   Amateurs who are registered with their local civil defense organization are permitted to participate in drills with the Radio Amateur Civil Emergency Service (RACES) for a maximum of **one hour per week**.

T2D11 How must you identify messages sent during a RACES drill?

A. As emergency messages
B. As amateur traffic
C. As official government messages
D. As drill or test messages

D   Any messages sent during a drill of the Radio Amateur Civil Emergency Service (RACES) must be identified as **drill** or **test messages**.

T2E   Voice communications and phonetics; SSB/CW weak signal operations; radioteleprinting; packet; special operations

T2E01 To make your call sign better understood when using voice transmissions, what should you do?

A. Use Standard International Phonetics for each letter of your call
B. Use any words that start with the same letters as your call sign for each letter of your call
C. Talk louder
D. Turn up your microphone gain

A   Use the **standard international phonetic alphabet** to spell your call sign when using voice transmissions. This will make it much easier for other operators to understand your call sign. For example, WR1B would say, "This is Whiskey Romeo One Bravo."

T2E02 What does the abbreviation "RTTY" stand for?

A. "Returning to you," meaning "your turn to transmit"
B. Radioteletype
C. A general call to all digital stations
D. Morse code practice over the air

B   RTTY is an abbreviation for **radioteletype**. The technical name for RTTY is narrow-bandwidth direct-printing telegraphy. The transmitting operator types a message on a keyboard and the receiving operator sees the message printed out on a teleprinter or computer monitor.

T2E03 What does "connected" mean in a packet-radio link?

A. A telephone link is working between two stations
B. A message has reached an amateur station for local delivery
C. A transmitting station is sending data to only one receiving station; it replies that the data is being received correctly
D. A transmitting and receiving station are using a digipeater, so no other contacts can take place until they are finished

C   Packet radio is an error-free digital (computer) communications system. When two packet radio stations are exchanging data, we say they are *connected*. The **transmitting station is sending data that is addressed only to one receiving station** and the **receiving station is sending acknowledgment replies** as the data is received correctly.

T2E04 What does "monitoring" mean on a packet-radio frequency?

A. The FCC is copying all messages
B. A member of the Amateur Auxiliary to the FCC's Compliance and Information Bureau is copying all messages
C. A receiving station is displaying all messages sent to it, and replying that the messages are being received correctly
D. A receiving station is displaying all messages on the frequency, and is not replying to any messages

D   Your packet radio station can receive signals from stations not transmitting directly to you. If your **station displays the information from all messages** it receives, but **does not respond** to the sending station, you are *monitoring* the packet activity.

T2E05 What is a digipeater?

    A. A packet-radio station that retransmits only data that is marked to be retransmitted

    B. A packet-radio station that retransmits any data that it receives

    C. A repeater that changes audio signals to digital data

    D. A repeater built using only digital electronics parts

A   A digital repeater (*digipeater*) is a **packet radio station** that **retransmits data between stations** when the proper command has been issued to ask it to serve as a relay point. Any packet radio station can serve as a digipeater.

T2E06 What does "network" mean in packet radio?

    A. A way of connecting terminal-node controllers by telephone so data can be sent over long distances

    B. A way of connecting packet-radio stations so data can be sent over long distances

    C. The wiring connections on a terminal-node controller board

    D. The programming in a terminal-node controller that rejects other callers if a station is already connected

B   Packet radio stations can be connected, or linked so data can be sent between stations over long distances. Such linked stations form a network.

T2E07 When should digital transmissions be used on 2-meter simplex voice frequencies?

    A. In between voice syllables

    B. Digital operations should be avoided on simplex voice frequencies

    C. Only in the evening

    D. At any time, so as to encourage the best use of the band

B   There are certain frequencies set aside for digital operations on the 2-meter band. The common packet "channels" are 145.01, 145.03, 145.05, 145.07 and 145.09 MHz. You can probably find one or two of these frequencies in your area that have most of the packet activity. For a direct connection with another local amateur you may wish to avoid those very busy frequencies. On the other hand, you should **avoid using the designated FM voice simplex frequencies** because packet and FM voice are not compatible modes.

T2E08 Which of the following modes of communication are NOT available to a Technician class operator?

A. CW and SSB on HF bands
B. Amateur television (ATV)
C. EME (Moon bounce)
D. VHF packet, CW and SSB

A   With a Technician license you can sample virtually every type of Amateur Radio operating activity. Even without passing a Morse code exam you can use CW on the VHF and UHF bands. You can also use single sideband (SSB) voice on those bands. There are segments of each band dedicated to CW operation as well as other "weak signal" operation such as SSB. You can sample amateur television (ATV), Earth-Moon-Earth (EME or moonbounce) communication, amateur satellites and packet radio. **The only type of operating that you can't experience is on the high-frequency (HF) bands.** There is certainly no reason to feel confined to FM voice on 2 meters.

T2E09 What speed should you use when answering a CQ call using RTTY?

A. Half the speed of the received signal
B. The same speed as the received signal
C. Twice the speed of the received signal
D. Any speed, since RTTY systems adjust to any signal speed

B   There are several sending speeds used for RTTY operation. Always be sure your system is set to **transmit at the same speed you are receiving** the other station. Normally your station won't copy another station properly unless you are both using the same sending speed, but some computerized systems are able to adjust to the speed the other station is using. If your system is like that, be sure it is also set to adjust the sending speed.

T2E10 When may you operate your amateur station aboard a commercial aircraft?

A. At any time

B. Only while the aircraft is not in flight

C. Only with the pilot's specific permission and not while the aircraft is operating under Instrument Flight Rules

D. Only if you have written permission from the commercial airline company and not during takeoff and landing

C   The FCC Rules specify that you may operate an amateur station aboard an aircraft, but the **installation must be approved by the pilot in command**. There are some other conditions to be met as well. For example, **you may not operate** your station while the aircraft is operating **under Instrument Flight Rules** (IFR), unless the station meets all Federal Aviation Agency (FAA) Rules. [97.11]

T2E11 When may you operate your amateur station somewhere in the US besides the address listed on your license?

A. Only during times of emergency

B. Only after giving proper notice to the FCC

C. During an emergency or an FCC-approved emergency practice

D. Whenever you want to

D   The FCC does not put any limitations on when or where in the US you may **operate your amateur station.** [97.5 (a)]

Radio-Wave Propagation

Your Technician license-exam will have 3 questions taken from the 3 groups of questions in this Radio-Wave Propagation subelement, T3A, T3B and T3C.

T3A Line of sight; reflection of VHF/UHF signals

T3A01 How are VHF signals propagated within the range of the visible horizon?

A. By sky wave
B. By line of sight
C. By plane wave
D. By geometric refraction

B   VHF signals usually travel directly from the transmitting antenna to the receiving antenna between stations that are within each other's visible horizon. This communication is possible because of **line-of-sight** signals.

T3A02 When a signal travels in a straight line from one antenna to another, what is this called?

    A. Line-of-sight propagation
    B. Straight line propagation
    C. Knife-edge diffraction
    D. Tunnel ducting

  A   Radio waves travel in a straight line from a transmitting antenna, but they can bend as they travel along the Earth's surface or into the ionosphere. Radio waves also reflect, or bounce off, objects in their paths. When radio waves travel directly from the transmitting antenna to the receiving antenna in a straight line, with no bending or reflection, we call it **line-of-sight propagation**.

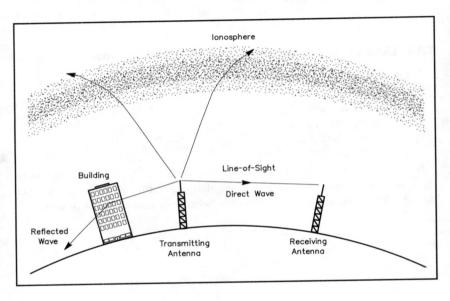

T3A03 How do VHF and UHF radio waves usually travel from a transmitting antenna to a receiving antenna?

    A. They bend through the ionosphere
    B. They go in a straight line
    C. They wander in any direction
    D. They move in a circle going either east or west from the transmitter

  B   The most common communications path for VHF and UHF signals is in a **straight line** from the transmitting antenna to the receiving antenna.

T3A04 What type of propagation usually occurs from one hand-held VHF transceiver to another nearby?

- A. Tunnel propagation
- B. Sky-wave propagation
- C. Line-of-sight propagation
- D. Auroral propagation

C   If you and a friend use 222-MHz hand-held radios to communicate directly with each other over a distance of a few miles, you are using **line-of-sight propagation**. This is the most common form of propagation when you are using VHF or UHF radios.

T3A05 What causes the ionosphere to form?

- A. Solar radiation ionizing the outer atmosphere
- B. Temperature changes ionizing the outer atmosphere
- C. Lightning ionizing the outer atmosphere
- D. Release of fluorocarbons into the atmosphere

A   Electromagnetic waves and particles from the sun hit the gas molecules in the Earth's upper atmosphere, knocking electrons free from the gas atoms and producing ions, or electrically charged particles. This **solar radiation produces an area of ionization.** This region is called the ionosphere.

T3A06 What type of solar radiation is most responsible for ionization in the outer atmosphere?

- A. Thermal
- B. Non-ionized particle
- C. Ultraviolet
- D. Microwave

C   **Ultraviolet radiation** from the sun is most responsible for the ionization in the outer atmosphere. (Ultraviolet radiation has a higher frequency — shorter wavelength — than visible light.)

T3A07 Which two daytime ionospheric regions combine into one region at night?

- A. E and F1
- B. D and E
- C. F1 and F2
- D. E1 and E2

C   The **F1 and F2 regions** of the ionosphere exist only during the daytime under normal conditions. They combine to form the F region at night.

T3A08 Which ionospheric region becomes one region at night, but separates into two separate regions during the day?

A. D
B. E
C. F
D. All of these choices

C   The F1 and F2 regions of the ionosphere exist only during the daytime, under normal conditions. They combine to form the **F region** at night.

T3A09 Ultraviolet solar radiation is most responsible for ionization in what part of the atmosphere?

A. Inner
B. Outer
C. All of these choices
D. None of these choices

B   Ultraviolet radiation from the sun produces charged molecules (ions) in the portion of the **outer atmosphere** called the ionosphere. These *ionized* particles allow high-frequency (HF) radio signals to travel large distances.

T3A10 What part of our atmosphere is formed by solar radiation ionizing the outer atmosphere?

A. Ionosphere
B. Troposphere
C. Ecosphere
D. Stratosphere

A   Ultraviolet radiation from the sun produces charged molecules (ions) in the portion of the outer atmosphere called the **ionosphere**. These ionized particles allow high-frequency (HF) radio signals to travel large distances.

T3A11 What can happen to VHF or UHF signals going towards a metal-framed building?

    A. They will go around the building
    B. They can be bent by the ionosphere
    C. They can be easily reflected by the building
    D. They are sometimes scattered in the ecosphere

    C   VHF and UHF signals are **easily reflected by buildings**, mountains and other objects in their paths. If there is something in the way that blocks a direct signal from your station to a friend's station, you might be able to bounce your signals off a building or mountain. In that case, you may have to point your beam (directional) antenna away from your friend's location and toward the building or mountain to produce the best signals.

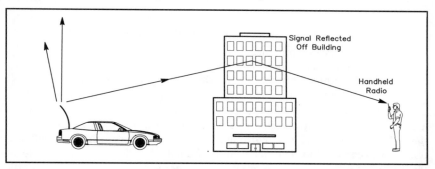

T3B   Tropospheric ducting or bending; amateur satellite and EME operations

T3B01 Ducting occurs in which region of the atmosphere?

    A. F2
    B. Ecosphere
    C. Troposphere
    D. Stratosphere

    C   The **troposphere** is a layer of the atmosphere that is below the ionosphere. When a warm air mass covers a cold air mass in the troposphere, a **duct** may form that traps radio waves in the cold air.

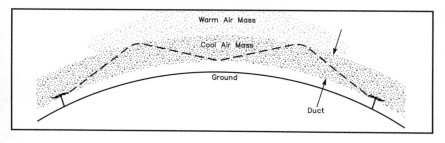

T3B02 What effect does tropospheric bending have on 2-meter radio waves?

    A. It lets you contact stations farther away
    B. It causes them to travel shorter distances
    C. It garbles the signal
    D. It reverses the sideband of the signal

A   Air masses with distinct differences in temperature that are next to each other can bend radio waves as they travel through the troposphere. This bending can extend the useful range of stations, **allowing contacts with stations farther away** than normal. This effect is especially noticeable on the 2-meter band and higher frequencies.

T3B03 What causes tropospheric ducting of radio waves?

    A. A very low pressure area
    B. An aurora to the north
    C. Lightning between the transmitting and receiving stations
    D. A temperature inversion

D   See the explanation to question T3B01. When a warm air mass covers a cold air mass in the troposphere, a **duct** may form that traps radio waves in the cold air. The weather condition that often forms such ducts is called a **temperature inversion**. This name comes from the fact that we normally expect the air temperature to become gradually cooler as the height above the Earth increases.

T3B04 What causes VHF radio waves to be propagated several hundred miles over oceans?

    A. A polar air mass
    B. A widespread temperature inversion
    C. An overcast of cirriform clouds
    D. A high-pressure zone

B   See the explanations to questions T3B01 and T3B03. **A widespread temperature inversion** can propagate VHF radio waves several hundred miles. Such temperature inversions usually occur over oceans, although they can also occur over land.

T3B05 In which of the following frequency ranges does tropospheric ducting most often occur?

A. UHF
B. MF
C. HF
D. VLF

A   Tropospheric ducting is the most common type of enhanced propagation for radio signals in the **UHF** range.

T3B06 What weather condition may cause tropospheric ducting?

A. A stable high-pressure system
B. An unstable low-pressure system
C. A series of low-pressure waves
D. Periods of heavy rainfall

A   VHF operators often pay attention to their local weather maps for help in predicting enhanced propagation such as through tropospheric ducting. If the map shows a **stable high-pressure system,** a temperature inversion is likely to occur, and that may produce some good conditions for ducting.

T3B07 How does the signal loss for a given path through the troposphere vary with frequency?

A. There is no relationship
B. The path loss decreases as the frequency increases
C. The path loss increases as the frequency increases
D. There is no path loss at all

C   All radio signals lose some strength as they travel through the troposphere. That **"path loss" increases for a given path as the signal frequency increases.**

T3B08 Why are high-gain antennas normally used for EME (moonbounce) communications?

    A. To reduce the scattering of the reflected signal as it returns to Earth

    B. To overcome the extreme path losses of this mode

    C. To reduce the effects of polarization changes in the received signal

    D. To overcome the high levels of solar noise at the receiver

B   The Moon's average distance from the Earth is 239,000 miles, and EME signals must travel twice that distance! Path losses are huge when compared to "local" VHF paths. Path loss refers to the total signal loss between the transmitting and receiving stations as compared to the total radiated signal energy. In addition, the Moon's surface is irregular and not a very efficient reflector of radio waves.

A typical EME station uses high-gain antennas and a high power amplifier to **overcome these extreme path losses**.

T3B09 Which of the following antenna systems would be the best choice for an EME (moonbounce) station?

    A. A single dipole antenna

    B. An isotropic antenna

    C. A ground-plane antenna

    D. A high-gain array of Yagi antennas

D   A typical EME station uses a high-gain antenna and a high-power amplifier. **A high-gain array of Yagi antennas** would be a good choice for a moonbounce station. You would not even want to try moonbounce with a single dipole or ground-plane antenna, no matter how much transmitter power you had!

EME operators really do use antennas like this (and larger) for moonbounce work. This photo shows the center four Yagis of an EME array for 432-MHz operation. You can see the elevation rotator, which rotates the array up and down vertically (to keep the antennas pointed at the moon). *Photo courtesy of Michael Katzmann, NV3Z.*

T3B10 When is it necessary to use a higher transmitter power level when conducting satellite communications?

A. When the satellite is at its perigee
B. When the satellite is low to the horizon
C. When the satellite is fully illuminated by the sun
D. When the satellite is near directly overhead

B　When a satellite is directly overhead, and you can point your antenna at the satellite, relatively low power is required. It is even possible to operate through some satellites using simple vertical antennas and low-power hand-held radios when the satellite is high in the sky.

When the satellite is near the horizon, however, you may need **more power** even if you can point your gain antenna directly at the bird. This is primarily because your signal must travel a longer distance through relatively dense air close to the Earth.

T3B11 Which of the following conditions must be met before two stations can conduct real-time communications through a satellite?

A. Both stations must use circularly polarized antennas
B. The satellite must be illuminated by the sun during the communications
C. The satellite must be in view of both stations simultaneously
D. Both stations must use high-gain antenna systems

C　Satellite communication uses line-of-sight propagation. Two amateurs can communicate through a satellite as long as **the satellite is in view of both stations** at the same time (**simultaneously**).

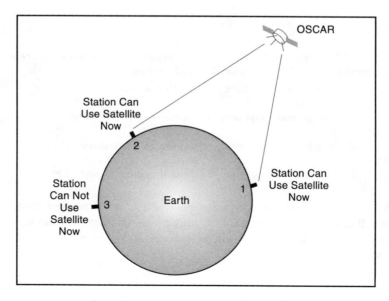

T3C Ionospheric propagation, causes and variation; maximum usable frequency; sporadic-E propagation; ground wave, HF propagation characteristics; sunspots and the sunspot cycle

T3C01 Which region of the ionosphere is mainly responsible for absorbing MF/HF radio signals during the daytime?

A. The F2 region
B. The F1 region
C. The E region
D. The D region

D The **D region** of the ionosphere exists only during the daytime, and it absorbs long-wavelength (low-frequency) radio signals that try to pass through it. Signals in the amateur 160, 80 and 40-meter (MF/HF) bands are most affected.

T3C02 If you are receiving a weak and distorted signal from a distant station on a frequency close to the maximum usable frequency, what type of propagation is probably occurring?

A. Ducting
B. Line-of-sight
C. Scatter
D. Ground-wave

C **Scatter** signals are often weak and distorted. If you hear such signals from a distant station on a frequency close to the maximum usable frequency, you are probably hearing scatter signals.

T3C03 In relation to sky-wave propagation, what does the term "maximum usable frequency" (MUF) mean?

A. The highest frequency signal that will reach its intended destination
B. The lowest frequency signal that will reach its intended destination
C. The highest frequency signal that is most absorbed by the ionosphere
D. The lowest frequency signal that is most absorbed by the ionosphere

A The maximum usable frequency (MUF) is the **highest-frequency signal** that is bent back to Earth to provide communications **between two specific locations**. The MUF for different propagation paths may be very different.

T3C04 When a signal travels along the surface of the Earth, what is this called?

A. Sky-wave propagation
B. Knife-edge diffraction
C. E-region propagation
D. Ground-wave propagation

D　Radio signals follow the surface of the Earth, even going over low hills in some cases, producing **ground-wave propagation**. Low-frequency, long-wavelength signals travel farthest along the ground; as far as 100 miles for stations near the low-frequency end of the standard AM broadcast band (the 540-kHz end). Higher-frequency, shorter-wavelength signals cannot be heard as far away. On the 80-meter Novice band, signals may travel up to about 70 miles by ground wave. At 28 MHz, in the amateur 10-meter band, you won't hear ground-wave signals from more than 10 to 15 miles away.

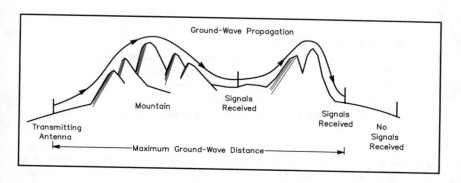

T3C05 When a signal is returned to Earth by the ionosphere, what is this called?

A. Sky-wave propagation
B. Earth-Moon-Earth propagation
C. Ground-wave propagation
D. Tropospheric propagation

A　Several things can happen to a radio signal that travels into the ionosphere. Some signals are absorbed by the electrically charged particles (ions) in the ionosphere, while others simply pass through the ionosphere and travel into space. (These are the signals you can use to communicate through a satellite, or with astronauts aboard the space shuttle, for example.) Still other signals are bent (refracted) by the ionosphere, and they return to Earth some distance away, producing **sky-wave propagation**.

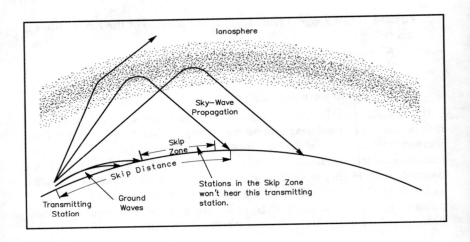

Ionosphere

Sky—Wave
Propagation

Skip Zone

Skip Distance

Stations in the Skip Zone
won't hear this transmitting
station.

Transmitting
Station

Ground
Waves

T3C06 What is a skip zone?

    A. An area covered by ground-wave propagation

    B. An area covered by sky-wave propagation

    C. An area that is too far away for ground-wave propagation, but too close for sky-wave propagation

    D. An area that is too far away for ground-wave or sky-wave propagation

    C   *Sky-wave*, or *skip propagation* has both a maximum range for a single hop, and a minimum range. The actual distances depend on the amount of ionization in the ionosphere, the frequency of the radio signals, and other effects. When the station you are trying to communicate with is beyond the range of ground-wave signals, but closer than the minimum skip distance (too close for sky-wave propagation), the station is in your **skip zone**.

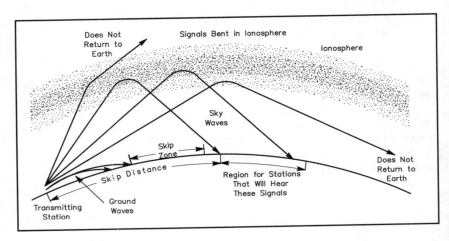

Does Not
Return to
Earth

Signals Bent in Ionosphere

Ionosphere

Sky
Waves

Skip Zone

Skip Distance

Does Not
Return to
Earth

Region for Stations
That Will Hear
These Signals

Transmitting
Station

Ground
Waves

T3C07 Which ionospheric region is closest to the Earth?

   A. The A region
   B. The D region
   C. The E region
   D. The F region

B   The **D region** of the ionosphere is only about 30 miles above the surface of the Earth. It is the closest region. When the regions of the ionosphere were named, scientists thought there may be other regions closer to the Earth, so they started naming them with the letter D. No closer regions were found, however.

T3C08 Which region of the ionosphere is mainly responsible for long-distance sky-wave radio communications?

   A. D region
   B. E region
   C. F1 region
   D. F2 region

D   Most long-distance sky-wave radio communications occurs because the radio signals are bent back toward the Earth by the **F2 region**. The F2 region is Frequently responsible for communications with Far-away places. Also see the explanation for question T3C07.

T3C09 Which of the ionospheric regions may split into two regions only during the daytime?

   A. Troposphere
   B. F
   C. Electrostatic
   D. D

B   There are three distinct regions of the ionosphere: the D region, the E region and the **F region**. During the day, the F region may split into two sub-regions, called the **F1** and the **F2** regions.

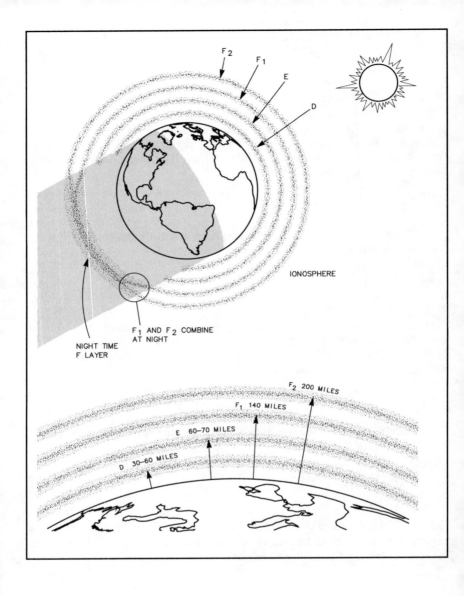

T3C10 How does the number of sunspots relate to the amount of ionization in the ionosphere?

    A. The more sunspots there are, the greater the ionization
    B. The more sunspots there are, the less the ionization
    C. Unless there are sunspots, the ionization is zero
    D. Sunspots do not affect the ionosphere

  A   Sunspots are grayish-black blotches on the sun's surface. **More sunspots usually mean more ionization** of the ionosphere. As a result, the higher-frequency bands tend to open for longer-distance communications when there are more sunspots.

T3C11 How long is an average sunspot cycle?

    A. 2 years
    B. 5 years
    C. 11 years
    D. 17 years

  C   The number and size of sunspots vary over approximately an **11-year cycle**. During a peak in sunspot activity you will often be able to communicate all over the world using a few watts of transmitter power on the 10-meter band. During the sunspot minimum you will have to move to lower-frequency bands like 40 and 80 meters for reliable worldwide communications.

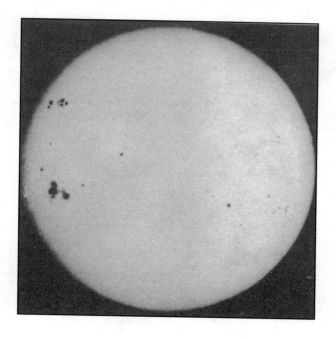

T4

Amateur Radio Practices

Four questions on your Technician exam will be from the Amateur Radio Practices subelement. Those questions will be taken from the 4 groups of questions printed in this chapter.

T4A Lightning protection and station grounding; safety interlocks, antenna installation safety procedures; dummy antennas

T4A01 How can an antenna system best be protected from lightning damage?

A. Install a balun at the antenna feed point
B. Install an RF choke in the antenna feed line
C. Ground all antennas when they are not in use
D. Install a fuse in the antenna feed line

    C   You should **ground all antenna feed lines** and other control cables **when you are not using your station**, to protect against lightning damage. You can also install lightning arrestors in the feed lines.

T4A02 How can amateur station equipment best be protected from lightning damage?

A. Use heavy insulation on the wiring
B. Never turn off the equipment
C. Disconnect the ground system from all radios
D. Disconnect all equipment from the power lines and antenna cables

    D   If you want to be certain that electronic equipment won't be damaged by a lightning storm, **unplug the equipment from the ac power line and disconnect it from any antenna cables**, rotator control cables, telephone lines and any other equipment. Don't rely on switches and fuses to protect your expensive equipment.

T4A03 For best protection from electrical shock, what should be grounded in an amateur station?

A. The power supply primary
B. All station equipment
C. The antenna feed line
D. The AC power mains

B   The metal chassis of **each piece of station equipment should be connected to ground**. This is easily done by running a short, heavy conductor from the equipment chassis to a common conductor, which then goes to the station ground.

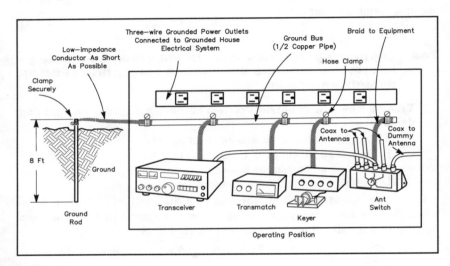

T4A04 Why would there be an interlock switch in a high-voltage power supply to turn off the power if its cabinet is opened?

A. To keep dangerous RF radiation from leaking out through an open cabinet
B. To keep dangerous RF radiation from coming in through an open cabinet
C. To turn the power supply off when it is not being used
D. To keep anyone opening the cabinet from getting shocked by dangerous high voltages

D   A safety interlock switch removes power when a door or cover is removed. Such a switch helps **protect you** and anyone else **from coming in contact with the dangerous high voltage levels** inside.

T4A05 Why should you wear a hard hat and safety glasses if you are on the ground helping someone work on an antenna tower?

A. So you won't be hurt if the tower should accidentally fall
B. To keep RF energy away from your head during antenna testing
C. To protect your head from something dropped from the tower
D. So someone passing by will know that work is being done on the tower and will stay away

C   Even a small bolt or a wrench can make a nasty dent when dropped from 50 or 100 feet in the air. If you are part of the "ground crew" on an antenna project, be sure to wear a hard hat **to protect your head**.

T4A06 What safety factors must you consider when using a bow and arrow or slingshot and weight to shoot an antenna-support line over a tree?

A. You must ensure that the line is strong enough to withstand the shock of shooting the weight
B. You must ensure that the arrow or weight has a safe flight path if the line breaks
C. You must ensure that the bow and arrow or slingshot is in good working condition
D. All of these choices are correct

D   A general rule for installing dipole and other wire antennas is "higher is better." Many amateurs find that a couple of tall trees make excellent end supports for such an antenna. Getting a line over one of the top limbs can be a challenge, however. A bow and arrow, with the line from a fishing reel attached to the arrow can easily launch the line over a tall tree. A slingshot, with a lead fishing sinker attached to the line from a fishing reel is another reliable method for somewhat shorter trees. If you are going to use one of these methods, however, you must consider several safety factors. Be sure **the line is strong enough** to withstand the shock of shooting the arrow or weight. **Check the flight path** to be sure it is safe down range. Be especially careful about where the projectile may go if the line breaks. You must also **check the bow and arrow or slingshot to ensure they are in good working order**. All the answer choices to this question are correct.

There are many ways to get an antenna support rope into a tree. These hams use a bow and arrow to shoot a lightweight fishing line over the desired branch. Then they attach the support rope to the fishing line and pull it up into the tree.

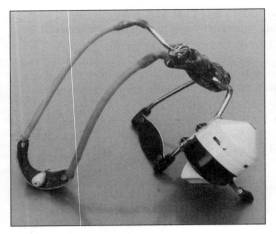

One method for getting an antenna support into a tree. Small hose clamps attach a casting reel to the wrist bracket of a slingshot. Monofilament fishing line attached to a 1-ounce sinker is easily shot over almost any tree. Remove the sinker and rewind the line for repeated shots. When you find a suitable path through the tree, use the fishing line to pull a heavier line over the tree.

T4A07 Which of the following is the best way to install your antenna in relation to overhead electric power lines?

- A. Always be sure your antenna wire is higher than the power line, and crosses it at a 90-degree angle
- B. Always be sure your antenna and feed line are well clear of any power lines
- C. Always be sure your antenna is lower than the power line, and crosses it at a small angle
- D. Only use vertical antennas within 100 feet of a power line

B   Never put your antenna or feed line under or over the top of electrical power lines. Always place antennas and feed lines far enough away from any power lines so there is no possibility that they can come in contact with each other. Your antennas must always be **well clear of any electric power lines**.

T4A08 What device is used in place of an antenna during transmitter tests so that no signal is radiated?

- A. An antenna matcher
- B. A dummy antenna
- C. A low-pass filter
- D. A decoupling resistor

B   Always use a **dummy antenna** when you are making transmitter adjustments. This prevents your signal from being radiated and causing unnecessary interference.

T4A09 Why would you use a dummy antenna?

A. For off-the-air transmitter testing
B. To reduce output power
C. To give comparative signal reports
D. To allow antenna tuning without causing interference

A   Use a dummy antenna **for off-the-air transmitter testing**. The dummy antenna absorbs the radio energy rather than sending it out over the air.

T4A10 What minimum rating should a dummy antenna have for use with a 100 watt single-sideband phone transmitter?

A. 100 watts continuous
B. 141 watts continuous
C. 175 watts continuous
D. 200 watts continuous

A   Your dummy antenna must be able to absorb all of your transmitter power. If you have a 100-watt transmitter, be sure your dummy antenna is rated for **100 watts continuous operation**. If you try putting too much power into the dummy antenna it will overheat, and may destroy the resistor.

T4A11 Would a 100 watt light bulb make a good dummy load for tuning a transceiver?

A. Yes; a light bulb behaves exactly like a dummy load
B. No; the impedance of the light bulb changes as the filament gets hot
C. No; the light bulb would act like an open circuit
D. No; the light bulb would act like a short circuit

B   Some older amateur literature suggested using a standard light bulb as a dummy antenna. A 100-watt light bulb would be used for a 100-watt transmitter, for example. The transmitter would be adjusted for maximum brightness of the bulb. While this was probably an acceptable dummy antenna for a tube-type transmitter, **it should not be used with a modern transistorized radio**. The **impedance of the light bulb changes** significantly **as the bulb heats up**.

T4B   Electrical wiring, including switch location, dangerous voltages and currents; SWR meaning and measurements; SWR meters

T4B01 Where should the green wire in a three-wire AC line cord be connected in a power supply?

A. To the fuse
B. To the "hot" side of the power switch
C. To the chassis
D. To the white wire

C   The ac power line cords used with most Amateur Radio transceivers and electrical appliances include three wires. One of the wires usually has *green* insulation to indicate that it is a *ground* wire, so the **green wire** should **connect to** the equipment **chassis**.

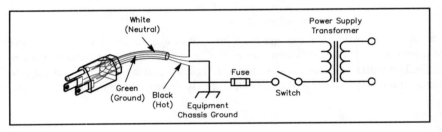

T4B02 What is the minimum voltage that is usually dangerous to humans?

A. 30 volts
B. 100 volts
C. 1000 volts
D. 2000 volts

A   Voltage is the force, or pressure that pushes an electrical current through a conductor. As little as **30 volts** is enough to push a dangerous current through your body.

T4B03 How much electrical current flowing through the human body will probably be fatal?

A. As little as 1/10 of an ampere
B. Approximately 10 amperes
C. More than 20 amperes
D. Current through the human body is never fatal

A   Small amounts of electrical current can cause dangerous shocks to the human body. As little as **1/10 of an ampere** (100 mA) can be fatal if the current flows through your heart.

T4B04 Which body organ can be fatally affected by a very small amount of electrical current?

A. The heart
B. The brain
C. The liver
D. The lungs

A   A small amount of electrical current can kill you if the current flows through your **heart**. The current can disrupt the normal beating of the heart.

T4B05 What does an SWR reading of less than 1.5:1 mean?

A. An impedance match that is too low
B. An impedance mismatch; something may be wrong with the antenna system
C. A fairly good impedance match
D. An antenna gain of 1.5

C   Any SWR reading of 2:1 or less is generally acceptable. A reading of 1.5:1 means the **impedance match is fairly good**.

T4B06 What does a very high SWR reading mean?

A. The antenna is the wrong length, or there may be an open or shorted connection somewhere in the feed line
B. The signals coming from the antenna are unusually strong, which means very good radio conditions
C. The transmitter is putting out more power than normal, showing that it is about to go bad
D. There is a large amount of solar radiation, which means very poor radio conditions

A   An **antenna that is not the correct length** may have a very high SWR. If the **feed line** becomes disconnected at the antenna — an **open circuit** — or if it becomes **short-circuited**, that will also cause a very high SWR reading.

T4B07 If an SWR reading at the low frequency end of an amateur band is 2.5:1, increasing to 5:1 at the high frequency end of the same band, what does this tell you about your 1/2-wavelength dipole antenna?

A. The antenna is broadbanded
B. The antenna is too long for operation on the band
C. The antenna is too short for operation on the band
D. The antenna is just right for operation on the band

B   An **antenna that is too long** will have a lower SWR reading at the low-frequency end of the band and the SWR will increase to a higher value at the high-frequency end of the band.

T4B08 If an SWR reading at the low frequency end of an amateur band is 5:1, decreasing to 2.5:1 at the high frequency end of the same band, what does this tell you about your 1/2-wavelength dipole antenna?

A. The antenna is broadbanded
B. The antenna is too long for operation on the band
C. The antenna is too short for operation on the band
D. The antenna is just right for operation on the band

C   An **antenna that is too short** will have a higher SWR reading at the low-frequency end of the band and the SWR will decrease to a lower value at the high-frequency end of the band.

T4B09 What instrument is used to measure the relative impedance match between an antenna and its feed line?

A. An ammeter
B. An ohmmeter
C. A voltmeter
D. An SWR meter

D   Standing wave ratio (SWR) indicates the relative impedance match between a transmitter and antenna. Use an **SWR meter**, or SWR bridge, to make this measurement. The SWR meter is sometimes called a reflectometer, because it measures the voltage of the RF energy going forward from the transmitter to the antenna, and the voltage reflected from the antenna back to the transmitter.

T4B10 If you use an SWR meter designed to operate on 3-30 MHz for VHF measurements, how accurate will its readings be?

A. They will not be accurate
B. They will be accurate enough to get by
C. If it properly calibrates to full scale in the set position, they may be accurate
D. They will be accurate providing the readings are multiplied by 4.5

C SWR meters are designed to work best over a specific frequency range. Since you are only looking for a relative indication of reflected power, however, you can **probably use** a meter designed to operate from 3 to 30-MHz for VHF measurements **if it properly calibrates to full scale in the set position.**

T4B11 What does an SWR reading of 1:1 mean?

A. An antenna for another frequency band is probably connected
B. The best impedance match has been attained
C. No power is going to the antenna
D. The SWR meter is broken

B An SWR reading of 1:1 means that the **antenna feed-point impedance is the same as the feed line characteristic impedance**, and **this is the best condition** you can hope for! The maximum amount of transmitter power is radiating from the antenna.

T4C   Meters and their placement in circuits, including volt, amp, multi, peak-reading and RF watt; ratings of fuses and switches

T4C01 How is a voltmeter usually connected to a circuit under test?

A. In series with the circuit
B. In parallel with the circuit
C. In quadrature with the circuit
D. In phase with the circuit

B   When you want to measure the voltage across a component or circuit, always connect the voltmeter in **parallel with the circuit**.

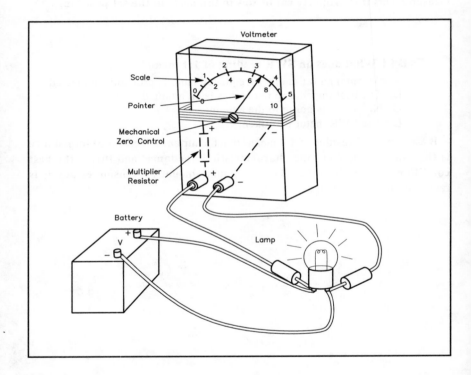

T4C02 How is an ammeter usually connected to a circuit under test?

    A. In series with the circuit
    B. In parallel with the circuit
    C. In quadrature with the circuit
    D. In phase with the circuit

A   When you want to measure the current through a component or circuit, always connect the ammeter in **series with the circuit**.

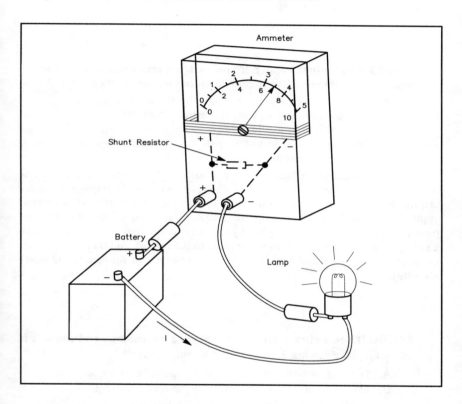

T4C03 Where should an RF wattmeter be connected for the most accurate readings of transmitter output power?

    A. At the transmitter output connector
    B. At the antenna feed point
    C. One-half wavelength from the transmitter output
    D. One-half wavelength from the antenna feed point

A   To measure your transmitter output power accurately, connect an RF wattmeter to the **transmitter output connector**, and then connect the antenna feed line to the wattmeter output terminal.

T4C04 For which measurements would you normally use a multimeter?

A. SWR and power
B. Resistance, capacitance and inductance
C. Resistance and reactance
D. Voltage, current and resistance

D   A multimeter is also often called a volt-ohm-milliammeter, or VOM. Such a meter can measure **voltage**, **current** and **resistance** with proper selection of the meter function and scale.

T4C05 What might happen if you switch a multimeter to measure resistance while you have it connected to measure voltage?

A. The multimeter would read half the actual voltage
B. It would probably destroy the meter circuitry
C. The multimeter would read twice the actual voltage
D. Nothing unusual would happen; the multimeter would measure the circuit's resistance

B   A multimeter includes a battery for resistance measurements. The battery supplies a small current through the resistor or other component you are testing. You should never try to measure resistance in a circuit that has power applied to it. Ideally you will only measure resistance for components that are removed from the circuit. If you have your meter connected to a circuit to make a voltage measurement. with power applied to the circuit, and then switch your meter to a resistance scale, you could **burn out the meter movement or other circuitry**.

T4C06 If you switch a multimeter to read microamps and connect it into a circuit drawing 5 amps, what might happen?

A. The multimeter would read half the actual current
B. The multimeter would read twice the actual current
C. It would probably destroy the meter circuitry
D. The multimeter would read a very small value of current

C   You must always be sure to select a current-measuring range that is greater than the circuit current you expect to find in the circuit. For example, if you were to set your multimeter to read a current in microamps and then connect it to a circuit that draws 5 amps, you will probably **burn out the meter, destroying the circuitry**. The safest approach is to always start with the highest meter setting, and then switch to a lower range as needed.

T4C07 At what line impedance do most RF watt meters usually operate?

A. 25 ohms
B. 50 ohms
C. 100 ohms
D. 300 ohms

B RF wattmeters can only measure power accurately if the transmitter output impedance and antenna feed-line impedance are the same as the wattmeter design impedance. Most RF wattmeters used by amateurs have a design impedance of **50 ohms**.

T4C08 What does a directional wattmeter measure?

A. Forward and reflected power
B. The directional pattern of an antenna
C. The energy used by a transmitter
D. Thermal heating in a load resistor

A A directional wattmeter includes a circuit to determine the **forward power**, flowing toward the antenna from the transmitter, and the **reflected power**, returning to the transmitter from the antenna.

T4C09 If a directional RF wattmeter reads 90 watts forward power and 10 watts reflected power, what is the actual transmitter output power?

A. 10 watts
B. 80 watts
C. 90 watts
D. 100 watts

B Power reflected back toward the transmitter from an antenna is reflected again when it gets to the transmitter. Since this power is now measured again as forward power, the wattmeter adds it to the actual transmitter output power. To determine the true transmitter output power, subtract the reflected power reading from the forward power reading.

Transmitter output power = forward power reading – reflected power reading

Transmitter output power = 90 watts – 10 watts = **80 watts**

T4C10 Why might you use a peak-reading RF wattmeter at your station?

- A. To make sure your transmitter's output power is not higher than that authorized by your license class
- B. To make sure your transmitter is not drawing too much power from the AC line
- C. To make sure all your transmitter's power is being radiated by your antenna
- D. To measure transmitter input and output power at the same time

A   The main reason for using a wattmeter in your amateur station is to **ensure that you are not exceeding the maximum power allowed by your license**. The FCC specifies this maximum power in terms of peak envelope power (PEP). The most accurate way to measure your power is with a peak-reading RF wattmeter.

T4C11 What could happen to your transceiver if you replace its blown 5 amp AC line fuse with a 30 amp fuse?

- A. The 30-amp fuse would better protect your transceiver from using too much current
- B. The transceiver would run cooler
- C. The transceiver could use more current than 5 amps and a fire could occur
- D. The transceiver would not be able to produce as much RF output

C   *Do not* put a larger fuse in an existing circuit. **The circuit may draw too much current, and a fire could result!** Find out what caused the fuse to blow and repair the problem. Replace the fuse only with another one of the same rating. For example, if you replaced a 5-amp fuse with one rated for 30 amps the **transceiver could draw too much current, causing the wires to overheat and even starting a fire**.

T4D   RFI and its complications, resolution and responsibility

T4D01 What is meant by receiver overload?

- A. Too much voltage from the power supply
- B. Too much current from the power supply
- C. Interference caused by strong signals from a nearby source
- D. Interference caused by turning the volume up too high

C   If a receiver does not have proper input filtering, **a strong signal from a nearby transmitter may cause interference** to the receiver even if the receiver operates on a frequency quite different from the transmitter frequency. Such receiver overload can often be cured by using additional receiver input filtering. Television interference is often the result of a TV receiver experiencing overload from a nearby transmitter.

T4D02 What is meant by harmonic radiation?

A. Unwanted signals at frequencies that are multiples of the fundamental (chosen) frequency
B. Unwanted signals that are combined with a 60-Hz hum
C. Unwanted signals caused by sympathetic vibrations from a nearby transmitter
D. Signals that cause skip propagation to occur

A   A harmonic signal is a **whole-number multiple of the fundamental, or chosen operating frequency.** If unwanted signals at 2, 3 or 4 times the fundamental frequency get to the antenna and are radiated, they can cause interference to receivers tuned to the harmonic frequencies.

T4D03 What type of filter might be connected to an amateur HF transmitter to cut down on harmonic radiation?

A. A key-click filter
B. A low-pass filter
C. A high-pass filter
D. A CW filter

B   A **low-pass filter** will block the harmonic signals at the transmitter output, but will allow the amateur HF signals to pass through to the antenna.

A low-pass filter. When connected in the coaxial cable feed line between an amateur transmitter and the antenna, a low-pass filter can reduce the strength of transmitted harmonics.

T4D04 If your neighbor reports television interference whenever you are transmitting from your amateur station, no matter what frequency band you use, what is probably the cause of the interference?

A. Too little transmitter harmonic suppression
B. Receiver VR tube discharge
C. Receiver overload
D. Incorrect antenna length

C   **Receiver overload,** or front-end overload is the most common type of interference caused by amateur transmitters. Having a high-pass filter installed on the TV antenna-input terminals is the best first step to solving this type of interference.

T4D05 If your neighbor reports television interference on one or two channels only when you are transmitting on the 15-meter band, what is probably the cause of the interference?

    A. Too much low-pass filtering on the transmitter
    B. De-ionization of the ionosphere near your neighbor's TV antenna
    C. TV receiver front-end overload
    D. Harmonic radiation from your transmitter

D   **Harmonics from your transmitter** can cause interference to a TV receiver. In this case only one or two VHF channels will likely be affected, and the interference will only occur when you are transmitting on one or two bands, such as the 10 or 15-meter band.

T4D06 What type of filter should be connected to a TV receiver as the first step in trying to prevent RF overload from an amateur HF station transmission?

    A. Low-pass
    B. High-pass
    C. Band pass
    D. Notch

B   A **high-pass filter** connected to the antenna input terminals of a TV receiver will block the signals from an amateur HF transmitter, while allowing the VHF and UHF TV signals through. This is a good first step to curing an interference problem caused by receiver (RF) overload.

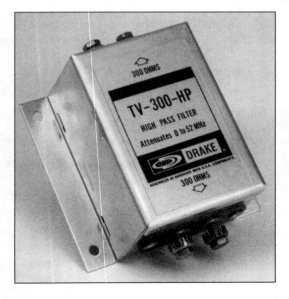

A high-pass filter can prevent fundamental energy from an amateur signal from entering a television set. This type of high-pass filter goes in the 300-ohm feed line that connects the television set with the antenna.

T4D07 What first step should be taken at a cable TV receiver when trying to prevent RF overload from an amateur HF station transmission?

A. Install a low-pass filter in the cable system transmission line
B. Tighten all connectors and inspect the cable system transmission line
C. Make sure the center conductor of the cable system transmission line is well grounded
D. Install a ceramic filter in the cable system transmission line

B The first step in trying to cure a receiver overload problem with a cable-TV receiver is to have the owner or a service technician **tighten all connectors** and **inspect the cable system transmission line**. Any loose connector or break in the transmission line can allow an amateur signal to leak into the line, causing interference to TV receivers.

T4D08 What effect might a break in a cable television transmission line have on amateur communications?

A. Cable lines are shielded and a break cannot affect amateur communications
B. Harmonic radiation from the TV receiver may cause the amateur transmitter to transmit off-frequency
C. TV interference may result when the amateur station is transmitting, or interference may occur to the amateur receiver
D. The broken cable may pick up very high voltages when the amateur station is transmitting

C Any loose connector or break in the transmission line can allow an amateur signal to leak into the line, **causing interference to TV receivers**. The first step in trying to cure an interference problem with a cable-TV receiver is to have the owner or a service technician tighten all connectors and inspect the cable system transmission line.

T4D09 If you are told that your amateur station is causing television interference, what should you do?

A. First make sure that your station is operating properly, and that it does not cause interference to your own television
B. Immediately turn off your transmitter and contact the nearest FCC office for assistance
C. Connect a high-pass filter to the transmitter output and a low-pass filter to the antenna-input terminals of the television
D. Continue operating normally, because you have no reason to worry about the interference

A   You should be aware that your Amateur Radio station has the potential to cause interference to a TV or other consumer electronics equipment. It *is* possible to operate your transmitter while others watch interference-free TV, however! Be sure your **station equipment** is properly designed and installed, and that it **is operating properly**. Take every step to **demonstrate** that your **transmitter does not interfere with the consumer electronics equipment in your own home**.

T4D10 If harmonic radiation from your transmitter is causing interference to television receivers in your neighborhood, who is responsible for taking care of the interference?

A. The owners of the television receivers are responsible
B. Both you and the owners of the television receivers share the responsibility
C. You alone are responsible, since your transmitter is causing the problem
D. The FCC must decide if you or the owners of the television receivers are responsible

C   **Harmonic interference must be cured at your transmitter**. As a licensed amateur, **you must take steps to ensure that all harmonics** from your transmitter **do not interfere** with other services. All harmonics generated by your transmitter must be attenuated well below the strength of the fundamental frequency.

T4D11 If signals from your transmitter are causing front-end overload in your neighbor's television receiver, who is responsible for taking care of the interference?

A. You alone are responsible, since your transmitter is causing the problem

B. Both you and the owner of the television receiver share the responsibility

C. The FCC must decide if you or the owner of the television receiver are responsible

D. The owner of the television receiver is responsible

D   There is nothing you can do to your transmitter to cure receiver overload interference. This is a fundamental problem with the receiving system, and the primary responsibility for curing the problem is with the **owner and the manufacturer**.

T5

Electrical Principles

Your Technician exam will will have 3 questions from this Electrical Principles subelement. Those questions will be taken from the 3 groups of questions printed in this chapter.

T5A   Metric prefixes, e.g. pico, nano, micro, milli, centi, kilo, mega, giga; concepts, units and measurement of current, voltage; concept of conductor and insulator; concept of open and short circuits

T5A01 If a dial marked in kilohertz shows a reading of 28450 kHz, what would it show if it were marked in hertz?

A. 284,500 Hz
B. 28,450,000 Hz
C. 284,500,000 Hz
D. 284,500,000,000 Hz

B   Use the chart to change from kilohertz (kHz) to hertz (Hz). The basic unit (U) on the chart represents hertz. Count three places to the right, and move the decimal point three places to the right to change 28450 kHz to **28,450,000 Hz.**

10^9	10^6	10^3	10^2	10^1	10^0	10^{-1}	10^{-2}	10^{-3}	10^{-6}	10^{-9}	10^{-12}
G • •	M • •	k	h	da	U	d	c	m • •	μ • •	n • •	p
giga	mega	kilo	hecto	deca	(unit)	deci	centi	milli	micro	nano	pico

T5A02 If an ammeter marked in amperes is used to measure a 3000-milliampere current, what reading would it show?

A. 0.003 amperes
B. 0.3 amperes
C. 3 amperes
D. 3,000,000 amperes

C   Use the chart to see how to change from milliamperes (mA) to amperes (A). The basic unit (U) on the chart represents amperes. Count three places to the left, and move the decimal point three places to the left to change 3000 mA to **3 A.**

T5A03 How many hertz are in a kilohertz?

- A. 10
- B. 100
- C. 1000
- D. 1,000,000

C   Use the chart with question T5A01 to change from kilohertz (kHz) to hertz (Hz). The basic unit (U) on the chart represents hertz. Count three places to the right, and move the decimal point three places to the right to change 1 kHz to **1000 Hz**.

T5A04 What is the basic unit of electric current?

- A. The volt
- B. The watt
- C. The ampere
- D. The ohm

C   The basic unit of electric current is called the **ampere**, abbreviated A. A current of 1 A represents 6.24×10^{18} electrons every second!

T5A05 Which instrument would you use to measure electric current?

- A. An ohmmeter
- B. A wavemeter
- C. A voltmeter
- D. An ammeter

D   An **ammeter** measures the current flowing through an electrical circuit.

T5A06 Which instrument would you use to measure electric potential or electromotive force?

- A. An ammeter
- B. A voltmeter
- C. A wavemeter
- D. An ohmmeter

B   A **voltmeter** measures the voltage or electric potential across an electrical circuit.

T5A07 What is the basic unit of electromotive force (EMF)?

A. The volt
B. The watt
C. The ampere
D. The ohm

A   We measure electromotive force, or voltage, in units of **volts**. This represents the pressure or force that pushes electrons through an electric circuit.

T5A08 What are three good electrical conductors?

A. Copper, gold, mica
B. Gold, silver, wood
C. Gold, silver, aluminum
D. Copper, aluminum, paper

C   Most metals conduct electricity, but nonmetals do not. **Gold, silver, aluminum** and **copper** are all good conductors.

T5A09 What are four good electrical insulators?

A. Glass, air, plastic, porcelain
B. Glass, wood, copper, porcelain
C. Paper, glass, air, aluminum
D. Plastic, rubber, wood, carbon

A   Materials that do not contain metal are often good insulators. **Glass, air, plastic, porcelain** and **rubber** are good insulators. Dry wood and paper can also be used as insulators.

T5A10 Which electrical circuit can have no current?

A. A closed circuit
B. A short circuit
C. An open circuit
D. A complete circuit

C   If an electrical circuit is **open** then no current can flow through the circuit. A *closed* or *complete* circuit has a complete path for the electrons to move through the circuit from the negative terminal to the positive terminal. A *short* circuit has an unwanted path for the electrons to flow from the negative terminal to the positive terminal.

T5A11 Which electrical circuit draws too much current?

A. An open circuit
B. A dead circuit
C. A closed circuit
D. A short circuit

D   A **short circuit** has an unwanted path for electrons to flow from the negative terminal to the positive terminal. A short circuit draws more current than is intended.

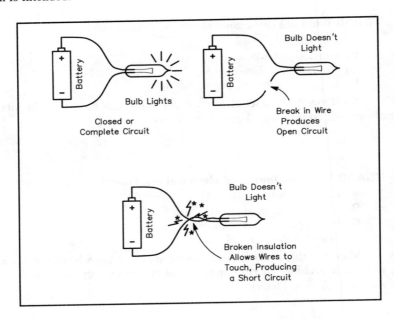

T5B   Concepts, units and calculation of resistance, inductance and capacitance values in series and parallel circuits

T5B01 What does resistance do in an electric circuit?

A. It stores energy in a magnetic field
B. It stores energy in an electric field
C. It provides electrons by a chemical reaction
D. It opposes the flow of electrons

D   Resistance **opposes the flow of electrons** in an electric circuit. We use resistance to control the amount of current that flows to various parts of a circuit.

T5B02 What is the definition of 1 ohm?

A. The reactance of a circuit in which a 1-microfarad capacitor is resonant at 1 MHz

B. The resistance of a circuit in which a 1-amp current flows when 1 volt is applied

C. The resistance of a circuit in which a 1-milliamp current flows when 1 volt is applied

D. The reactance of a circuit in which a 1-millihenry inductor is resonant at 1 MHz

B   The ohm is the basic unit used to measure circuit resistance. It is named for Georg Simon Ohm, a German physicist and mathematician who discovered the relationship between circuit voltage, current and resistance we call Ohm's Law. **A circuit that has a current of 1 amp when a voltage of 1 volt is applied has a resistance of 1 ohm.**

T5B03 What is the basic unit of resistance?

A. The farad

B. The watt

C. The ohm

D. The resistor

C   The **ohm** is the **basic unit** used to measure **circuit resistance**. It is named for Georg Simon Ohm, a German physicist and mathematician who discovered the relationship between circuit voltage, current and resistance we call Ohm's Law.

T5B04 What is one reason resistors are used in electronic circuits?

A. To block the flow of direct current while allowing alternating current to pass

B. To block the flow of alternating current while allowing direct current to pass

C. To increase the voltage of the circuit

D. To control the amount of current that flows for a particular applied voltage

D   Resistance **opposes the flow of electrons** in an electric circuit. We use resistance to **control the amount of current** that flows to various parts of a circuit **for a particular applied voltage**.

T5B05 What is the ability to store energy in a magnetic field called?

    A. Admittance
    B. Capacitance
    C. Resistance
    D. Inductance

D   Coils, or inductors, store electrical energy in the form of a magnetic field around the coil when current flows through the coil. This ability to store energy in a magnetic field is called **inductance**.

T5B06 What is one reason inductors are used in electronic circuits?

    A. To block the flow of direct current while allowing alternating current to pass
    B. To reduce the flow of AC while allowing DC to pass freely
    C. To change the time constant of the applied voltage
    D. To change alternating current to direct current

B   Inductors play several important roles in electronic circuits. When an ac signal is applied to the inductor, the inductance will act to oppose or **reduce the flow of ac**. When a dc signal is applied to an inductor, however, the inductance will have little effect on the current, so **dc is allowed to pass freely**.

T5B07 What is the ability to store energy in an electric field called?

    A. Inductance
    B. Resistance
    C. Tolerance
    D. Capacitance

D   Capacitors store electrical energy in the form of an electric field when voltage is applied across the capacitor. This ability to store energy in an electric field is called **capacitance**.

T5B08 What is one reason capacitors are used in electronic circuits?

    A. To block the flow of direct current while allowing alternating current to pass

    B. To block the flow of alternating current while allowing direct current to pass

    C. To change the time constant of the applied voltage

    D. To change alternating current to direct current

    A   Capacitors play several important roles in electronic circuits. When a dc signal is applied to the capacitor, the capacitor will charge the applied voltage and then no more current can flow in that circuit. So capacitors **block the flow of direct current**. When an ac signal is applied to a capacitor, however, the capacitor will alternately charge and discharge as the current reverses direction. A capacitor will have little effect on the current, so **alternating current is allowed to pass freely**.

T5B09 If two resistors are connected in series, what is their total resistance?

    A. The difference between the individual resistor values

    B. Always less than the value of either resistor

    C. The product of the individual resistor values

    D. The sum of the individual resistor values

    D   When resistors are connected **in series** their **resistance values add**. If two equal-value resistors are connected in series the total resistance will be **twice** the value of either resistor alone.

T5B10 If two equal-value inductors are connected in parallel, what is their total inductance?

    A. Half the value of one inductor

    B. Twice the value of one inductor

    C. The same as the value of either inductor

    D. The value of one inductor times the value of the other

    A   When inductors are connected in parallel the inductance of the combination will always be less than either value alone. To find the inductance of two inductors connected in parallel, use the following equation.

$$L_{TOTAL} = \frac{L_1 \times L_2}{L_1 + L_2}$$

If two equal-value inductors are connected in parallel the total value will be **half the inductance of either value alone**.

T5B11 If two equal-value capacitors are connected in series, what is their total capacitance?

A. Twice the value of one capacitor
B. The same as the value of either capacitor
C. Half the value of either capacitor
D. The value of one capacitor times the value of the other

C   When capacitors are connected in series the capacitance of the combination will always be less than either value alone. To find the capacitance of two capacitors connected in series, use the following equation. If two equal-value capacitors are connected in series the total value will be **half the capacitance of either value alone.**

T5C   Ohm's law (any calculations will be kept to a very low level - no fractions or decimals) and the concepts of energy and power, and; concepts of frequency, including AC vs. DC, frequency units, and wavelength

T5C01 How is the current in a DC circuit directly calculated when the voltage and resistance are known?

A. I = R × E [current equals resistance multiplied by voltage]
B. I = R / E [current equals resistance divided by voltage]
C. I = E / R [current equals voltage divided by resistance]
D. I = E / P [current equals voltage divided by power]

C   Ohm's Law tells us that the **current (I)** through a circuit is **equal** to the **voltage (E)** across the circuit **divided by** the circuit **resistance (R)**. Use the Ohm's Law Circle to find the correct form of the Ohm's Law equation. Cover the letter representing the quantity you want to find and notice the positions of the other two letters. Since we want to find current in this question, you will see that the E is above the R, representing a fraction with **E divided by R**.

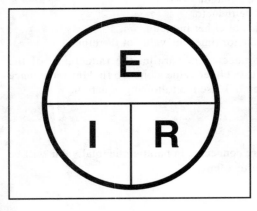

This simple diagram will help you remember the Ohm's Law relationships. To find any quantity if you know the other two, simply cover the unknown quantity with your hand or a piece of paper. The positions of the remaining two symbols show if you have to multiply (when they are side by side) or divide (when they appear one over the other as a fraction).

T5C02 How is the resistance in a DC circuit calculated when the voltage and current are known?

   A. R = I / E [resistance equals current divided by voltage]
   B. R = E / I [resistance equals voltage divided by current]
   C. R = I × E [resistance equals current multiplied by voltage]
   D. R = P / E [resistance equals power divided by voltage]

B   Ohm's Law tells us that the circuit **resistance (R)** is **equal** to the **voltage (E)** across the circuit divided by the **current (I)** through the circuit. Use the Ohm's Law Circle in question T5B01 to find the correct form of the Ohm's Law equation. Cover the letter representing the quantity you want to find and notice the positions of the other two letters. Since we want to find resistance in this question, you will see that the E is above the I, representing a fraction with **E divided by I.**

T5C03 How is the voltage in a DC circuit directly calculated when the current and resistance are known?

   A. E = I / R [voltage equals current divided by resistance]
   B. E = R / I [voltage equals resistance divided by current]
   C. E = I × R [voltage equals current multiplied by resistance]
   D. E = I / P [voltage equals current divided by power]

C   Ohm's Law tells us that the **voltage (E)** across a circuit is **equal** to the **current (I)** through the circuit multiplied by the circuit **resistance (R)**. Use the Ohm's Law Circle in question T5C01 to find the correct form of the Ohm's Law equation. Cover the letter representing the quantity you want to find and notice the positions of the other two letters. Since we want to find voltage in this question, you will see that the I and the R are side by side, so you multiply **I times R.**

$$E = I \times R$$

T5C04 If a current of 2 amperes flows through a 50-ohm resistor, what is the voltage across the resistor?

   A. 25 volts
   B. 52 volts
   C. 100 volts
   D. 200 volts

C   This is an Ohm's Law question. Find the correct form of the Ohm's Law equation using the Ohm's Law Circle. Cover the letter representing the quantity you want to find and notice the positions of the other two letters. Since we want to find voltage in this question, you will see that the I and R are side by side, so you multiply them.

$E = I \times R$ $E = 2 A \times 50 \Omega = 100 V$

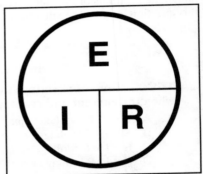

On your calculator enter this as: **ON 2 × 50 =** and read the answer as **100**.

Notice that we use the letter A to represent amperes, the unit for current. The letter V represents volts, the unit for voltage and the Greek capital letter omega (Ω) to represent ohms, the unit for resistance.

T5C05 If a 100-ohm resistor is connected to 200 volts, what is the current through the resistor?

   A. 1 ampere
   B. 2 amperes
   C. 300 amperes
   D. 20,000 amperes

B   See the explanation for question T5C04. This is an Ohm's Law question. Find the correct form of the Ohm's Law equation using the Ohm's Law Circle. Cover the letter representing the quantity you want to find and notice the positions of the other two letters. Since we want to find current in this question, you will see that the E is above the R, so you divide them.

$I = E / R$ $I = 200 V / 100 \Omega = 2 A$

On your calculator enter this as: **ON 200 ÷ 100 =** and read the answer as **2**.

T5C06 If a current of 3 amperes flows through a resistor connected to 90 volts, what is the resistance?

A. 3 ohms
B. 30 ohms
C. 93 ohms
D. 270 ohms

B See the explanation for question T5C04. This is an Ohm's Law question. Find the correct form of the Ohm's Law equation using the Ohm's Law Circle. Cover the letter representing the quantity you want to find and notice the positions of the other two letters. Since we want to find resistance in this question, you will see that the E is above the I, so you divide them.

$$R = E / I \qquad R = 90 \text{ V} / 3 \text{ A} = 30 \ \Omega$$

On your calculator enter this as: **ON 90 ÷ 3** = and read the answer as **30**.

T5C07 What term describes how fast electrical energy is used?

A. Resistance
B. Current
C. Power
D. Voltage

C **Power** describes how fast a circuit or electric appliance uses electrical energy.

T5C08 What is the basic unit of electrical power?

A. The ohm
B. The watt
C. The volt
D. The ampere

B We measure electrical power in **watts**, abbreviated W.

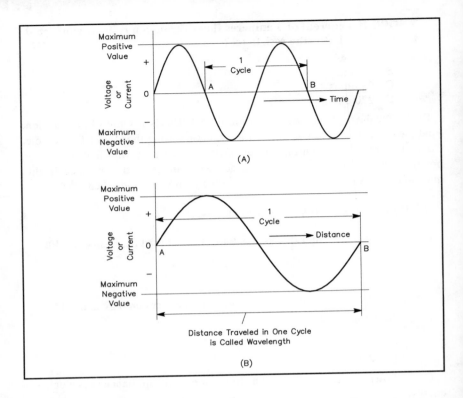

(A)

Distance Traveled in One Cycle
is Called Wavelength

(B)

T5C09 What happens to a signal's wavelength as its frequency increases?

A. It gets shorter
B. It gets longer
C. It stays the same
D. It disappears

A   The wavelength of an electrical signal **gets shorter** as the frequency increases. This is because electrical signals (such as radio waves) travel at the speed of light. So if the signal changes direction at a higher rate, it won't travel as far during one cycle. See the drawing with question T5C09.

T5C10 What is the name of a current that flows back and forth, first in one direction, then in the opposite direction?

A. An alternating current
B. A direct current
C. A rough current
D. A steady state current

A   An **alternating current** flows first in one direction and then in the opposite direction. The current changes direction because the polarity of the voltage source connected to the circuit changes direction.

T5C11 What is the name of a current that flows only in one direction?

- A. An alternating current
- B. A direct current
- C. A normal current
- D. A smooth current

B A **direct current** flows in one direction, from the negative terminal to the positive terminal. A battery is a good source of direct current.

Circuit Components

There will be 2 questions from the Circuit Components subelement on your exam. Those questions will come from the 2 groups of questions printed in this chapter.

T6A Electrical function and/or schematic representation of resistor, switch, fuse, or battery; resistor construction types, variable and fixed, color code, power ratings, schematic symbols

T6A01 What does a variable resistor or potentiometer do?

A. Its resistance changes when AC is applied to it
B. It transforms a variable voltage into a constant voltage
C. Its resistance changes when its slide or contact is moved
D. Its resistance changes when it is heated

C   Just as the name implies, a variable resistor allows you to **change**, or vary, the **resistance value**. (The name potentiometer implies that you can use this device to adjust the electrical *potential*, or voltage applied to a circuit.) Variable resistors use a **slide**, or movable **contact** point moving along a resistance element to change the resistance.

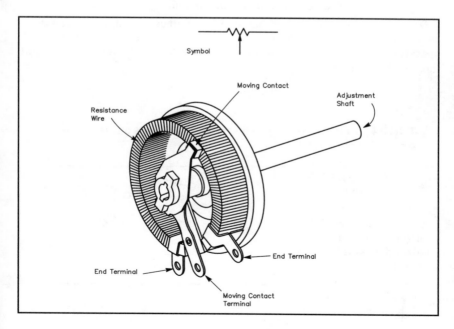

T6A02 Which symbol of Figure T6-1 represents a fixed resistor?

    A. Symbol 2
    B. Symbol 3
    C. Symbol 4
    D. Symbol 5

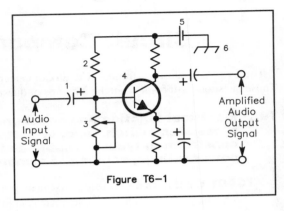

Figure T6–1

A   The circuit shown in Figure T6-1 is a simple common-emitter NPN bipolar transistor audio amplifier. Symbol 1 is an electrolytic capacitor, **symbol 2 is a fixed resistor**, symbol 3 is a variable resistor, symbol 4 is the NPN bipolar transistor, symbol 5 is a battery and symbol 6 is a chassis ground.

T6A03 Why would you use a double-pole, single-throw switch?

    A. To switch one input to one output
    B. To switch one input to either of two outputs
    C. To switch two inputs at the same time, one input to either of two outputs, and the other input to either of two outputs
    D. To switch two inputs at the same time, one input to one output, and the other input to the other output

D   A double-*pole* switch has two input lines. A single-*throw* switch connects each input to its own output line. A double-pole, single-throw switch **operates two circuits at the same time by connecting each input to the output line for that circuit**.

T6A04 In Figure N6-2, which symbol represents a single-pole, single-throw switch?

    A. Symbol 1
    B. Symbol 2
    C. Symbol 3
    D. Symbol 4

Figure N6–2

A   In Figure N6-2, **symbol 1** represents a single-pole, single-throw switch. This is a simple ON/OFF switch for one circuit.

To protect your equipment against being zapped, install fuses on both the positive and negative leads, close to the battery terminals.

T6A05 Why would you use a fuse?

A. To create a short circuit when there is too much current in a circuit
B. To change direct current into alternating current
C. To change alternating current into direct current
D. To create an open circuit when there is too much current in a circuit

D   To protect against unexpected short circuits and other problems, most electronic equipment includes one or more fuses. A fuse is simply a device made of metal (usually inside a glass enclosure) that will heat up and melt when a certain amount of current flows through it. When the fuse melts (or blows) it **creates an open circuit**, stopping the current.

T6A06 In Figure N6-1, which symbol represents a fuse?

A. Symbol 1
B. Symbol 3
C. Symbol 5
D. Symbol 7

A   In Figure N6-1, **symbol 1** represents a fuse. With this symbol you can imagine that the curved line between the dots represents the metal fuse element, which will melt if too much current flows through. That would then leave an open circuit between the dots.

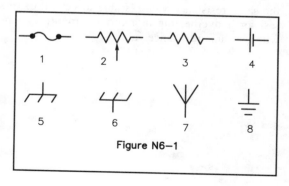

Figure N6—1

T6A07 Which of these components has a positive and a negative side?

    A. A battery
    B. A potentiometer
    C. A fuse
    D. A resistor

  A   Of the electrical components listed here, only a **battery** has a positive and a negative side. A battery supplies the voltage or electrical pressure to push electrons through a circuit from the negative terminal to the positive terminal.

T6A08 In Figure N6-1, which symbol represents a single-cell battery?

    A. Symbol 7
    B. Symbol 5
    C. Symbol 1
    D. Symbol 4

  D   In Figure N6-1, **symbol 4** represents a battery. The positive side is always shown with the longer line and the negative side with the shorter line. Sometimes + and – signs are included with the symbol, but they aren't really necessary.

T6A09 Why would a large size resistor be used instead of a smaller one of the same resistance value?

    A. For better response time
    B. For a higher current gain
    C. For greater power dissipation
    D. For less impedance in the circuit

  C   A resistor that is physically larger than a smaller-sized resistor with the same resistance value is able to get rid of (**dissipate**) **more power**. If a resistor is overheating in a certain circuit, try replacing it with a larger unit with the same resistance value.

T6A10 What do the first three color bands on a resistor indicate?

    A. The value of the resistor in ohms
    B. The resistance tolerance in percent
    C. The power rating in watts
    D. The resistance material

    A   Most film and carbon-composition **resistors use a color code** to show the resistance value. If the resistor has colored stripes, or bands, around the body, use the color code to determine the value. The first and second color bands on a resistor (starting with the band closest to one end) indicate the first two digits of the resistance value. The third color band indicates the multiplier, or number of zeros to be added after the two digits. A resistor whose first three color bands are yellow, violet and red, for example, is a 4700-Ω resistor.

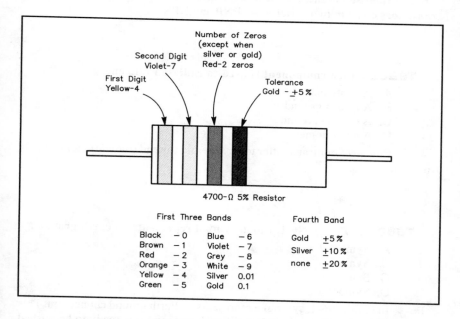

T6A11 Which tolerance rating would indicate a high-precision resistor?

    A. 0.1%
    B. 5%
    C. 10%
    D. 20%

    A   Resistors with smaller tolerance ratings have better precision, which means the actual resistor value is closer to the marked resistance. Common tolerance ratings are 5%, 10% and 20% values, but you can buy 1% and 0.1% tolerance resistors from most suppliers. A resistor **tolerance of 0.1% is a precision resistor**, and will generally cost more than a 20% or 10% unit.

T6B Electrical function and/or schematic representation of a ground, antenna, inductor, capacitor, transistor, integrated circuit; construction of variable and fixed inductors and capacitors; factors affecting inductance and capacitance

T6B01 Which component can amplify a small signal using low voltages?

A. A PNP transistor
B. A variable resistor
C. An electrolytic capacitor
D. A multiple-cell battery

A Transistors generally use low voltages to amplify small signals. **Bipolar transistors** come in two varieties — **PNP** and NPN.

T6B02 Which component is used to radiate radio energy?

A. An antenna
B. An earth ground
C. A chassis ground
D. A potentiometer

A Connect your transmitter to an **antenna** to send (radiate) radio signals from your station.

T6B03 In Figure N6-1, which symbol represents an earth ground?

A. Symbol 2
B. Symbol 5
C. Symbol 6
D. Symbol 8

D In Figure N6-1, **symbol 8** represents an Earth ground connection. This symbol looks a bit like a garden spade or shovel, and it is ready to be pushed into the Earth. You should connect the chassis or cabinet of each piece of radio equipment in your station to a good Earth ground.

T6B04 In Figure N6-1, which symbol represents an antenna?

A. Symbol 2
B. Symbol 3
C. Symbol 6
D. Symbol 7

D   In Figure N6-1, **symbol 7** represents an antenna. Imagine your radio signals leaving the outstretched arms of this symbol and traveling to some distant location.

T6B05 In Figure N6-3, which symbol represents an NPN transistor?

A. Symbol 1
B. Symbol 2
C. Symbol 3
D. Symbol 4

D   In Figure N6-3, **symbol 4** represents an NPN transistor. The three sections of a bipolar transistor are called the *emitter*, *base* and *collector*. Notice that the arrow on the emitter terminal is *N*ot *P*ointing i*N* on the NPN transistor symbol.

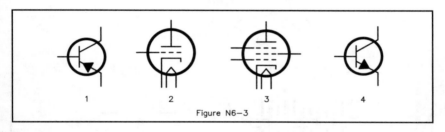

Figure N6-3

T6B06 Which symbol of Figure T6-2 represents a fixed-value capacitor?

A. Symbol 1
B. Symbol 2
C. Symbol 3
D. Symbol 4

A   Symbols 1 and 3 represent capacitors and symbols 2 and 4 represent inductors. Whenever you see a schematic symbol with an arrow through the symbol you can be sure that component is adjustable in some way. So **symbol 1 represents a fixed capacitor**, while symbol 3 is a variable capacitor.

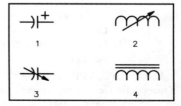

T6B07 In Figure T6-2, which symbol represents a variable capacitor?

A. Symbol 1
B. Symbol 2
C. Symbol 3
D. Symbol 4

C Symbols 1 and 3 represent capacitors and symbols 2 and 4 represent inductors. Whenever you see a schematic symbol with an arrow through the symbol you can be sure that component is adjustable in some way. So **symbol 3 represents a variable capacitor**, while symbol 1 is a fixed-value capacitor.

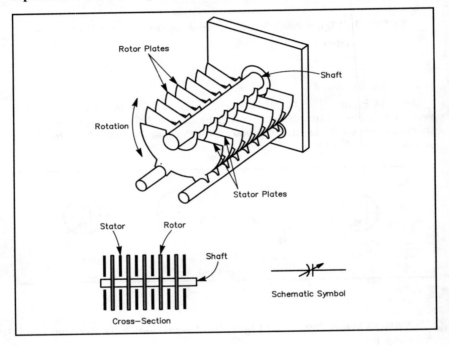

T6B08 What does an inductor do?

A. It stores energy electrostatically and opposes a change in voltage
B. It stores energy electrochemically and opposes a change in current
C. It stores energy electromagnetically and opposes a change in current
D. It stores energy electromechanically and opposes a change in voltage

C An inductor is a coil of wire that stores electrical **energy** in the form of a **magnetic field** when current flows through the wire. When an electric current produces a magnetic field, we have formed an **electromagnet. Inductors oppose any change in current** through the coil.

T6B09 As an iron core is inserted in a coil, what happens to the coil's inductance?

A. It increases
B. It decreases
C. It stays the same
D. It disappears

A   Iron is a magnetic material. If an iron core is inserted into the turns of wire of a coil, the inductance value will **increase**.

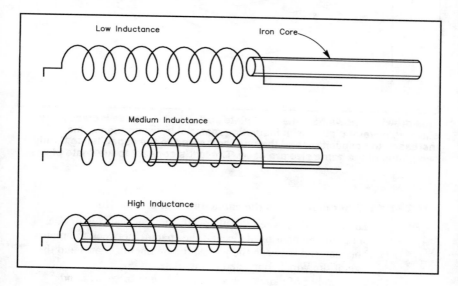

Low Inductance

Iron Core

Medium Inductance

High Inductance

T6B10 What does a capacitor do?

A. It stores energy electrochemically and opposes a change in current
B. It stores energy electrostatically and opposes a change in voltage
C. It stores energy electromagnetically and opposes a change in current
D. It stores energy electromechanically and opposes a change in voltage

B   A capacitor stores an electrical charge in the form of an **electrostatic field**, and opposes any change in the **voltage** applied across the capacitor plates.

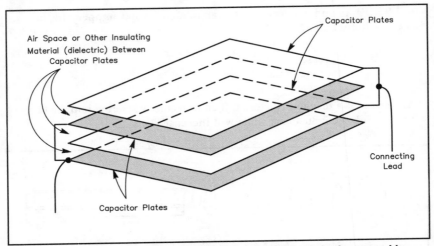

Air Space or Other Insulating Material (dielectric) Between Capacitor Plates

Capacitor Plates

Connecting Lead

Capacitor Plates

This drawing shows how the total plate surface area can be increased by stacking several plates, with insulating material between them. This increases the capacitance of the unit, compared to the capacitance if only two plates of the same size are used. The leads connect to alternate plates.

T6B11 What determines the capacitance of a capacitor?

A. The material between the plates, the area of one side of one plate, the number of plates and the spacing between the plates

B. The material between the plates, the number of plates and the size of the wires connected to the plates

C. The number of plates, the spacing between the plates and whether the dielectric material is N type or P type

D. The material between the plates, the area of one plate, the number of plates and the material used for the protective coating

A Several physical properties help determine the capacitance value of a capacitor. Among these are the **insulating** (dielectric) **material** between the plates, the **area** of one side of one plate, the **number** of plates and the **spacing** between the plates.

Practical Circuits

Your Technician written exam will have 2 questions from the Practical Circuits subelement. There are 2 groups of questions for this subelement.

T7A Functional layout of station components including transmitter, transceiver, receiver, power supply, antenna, antenna switch, antenna feed line, impedance-matching device, SWR meter; station layout and accessories for radiotelephone, radioteleprinter (RTTY) or packet

T7A01 What would you connect to your transceiver if you wanted to switch it between several antennas?
A. A terminal-node switch
B. An antenna switch
C. A telegraph key switch
D. A high-pass filter

B An **antenna switch** selects between several antennas, so you can choose the proper antenna to use with your transceiver on each band.

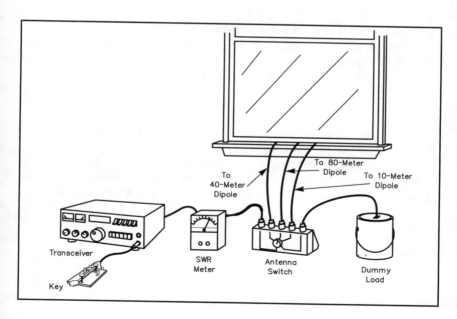

T7A02 What connects your transceiver to your antenna?

    A. A dummy load
    B. A ground wire
    C. The power cord
    D. A feed line

D   Your amateur transceiver is located in your "shack," which is usually a room or part of a room in your house. The antenna, which radiates your signals, usually works best located high in the air, and away from buildings and other structures. To connect the transceiver to the antenna, you will use a **feed line**. Coaxial cable is one popular type of feed line. Various types of parallel-conductor lines are also popular.

T7A03 If your mobile transceiver works in your car but not in your home, what should you check first?

    A. The power supply
    B. The speaker
    C. The microphone
    D. The SWR meter

A   Many radios operate from the 12-volt dc electrical system in a car. A power supply converts the 120 volt ac household electrical system to 12 volts dc. If your radio works in your car but not in your house check the **power supply** first.

T7A04 What does an antenna tuner do?

    A. It matches a transceiver output impedance to the antenna system impedance
    B. It helps a receiver automatically tune in stations that are far away
    C. It switches an antenna system to a transceiver when sending, and to a receiver when listening
    D. It switches a transceiver between different kinds of antennas connected to one feed line

A   Most modern amateur transceivers require an impedance close to 50 ohms at the antenna connector. If your antenna system impedance is much different from that, you can use an antenna tuner to **match the impedances**.

T7A05 In Figure N7-1, if block 1 is a transceiver and block 3 is a dummy antenna, what is block 2?

- A. A terminal-node switch
- B. An antenna switch
- C. A telegraph key switch
- D. A high-pass filter

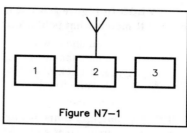

Figure N7-1

B An **antenna switch** allows you to select one of several antennas to be connected to your transceiver. A dummy antenna is useful for transmitter testing and adjustments.

T7A06 In Figure N7-1, if block 1 is a transceiver and block 2 is an antenna switch, what is block 3?

- A. A terminal-node switch
- B. An SWR meter
- C. A telegraph key switch
- D. A dummy antenna

D An antenna switch allows you to select one of several antennas to be connected to your transceiver. A **dummy antenna** is useful for transmitter testing and adjustments.

T7A07 In Figure N7-2, if block 1 is a transceiver and block 3 is an antenna switch, what is block 2?

- A. A terminal-node switch
- B. A dipole antenna
- C. An SWR meter
- D. A high-pass filter

C An antenna switch allows you to select one of several antennas to be connected to your transceiver. An **SWR meter** is often included between the transceiver and the antenna switch.

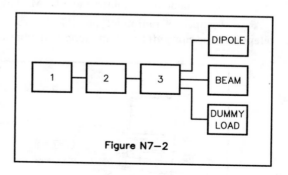

Figure N7-2

T7A08 In Figure N7-3, if block 1 is a transceiver and block 2 is an SWR meter, what is block 3?

    A. An antenna switch
    B. An antenna tuner
    C. A key-click filter
    D. A terminal-node controller

B   An **antenna tuner** is often placed after a transceiver and SWR meter, but before the antenna, to match the antenna system impedance to the transceiver 50-ohm output.

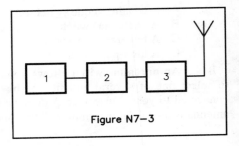

Figure N7-3

T7A09 What would you connect to a transceiver for voice operation?

    A. A splatter filter
    B. A terminal-voice controller
    C. A receiver audio filter
    D. A microphone

D   You will have to connect a **microphone** to your transceiver for voice operation.

T7A10 What would you connect to a transceiver for RTTY operation?

    A. A modem and a teleprinter or computer system
    B. A computer, a printer and a RTTY refresh unit
    C. A data-inverter controller
    D. A modem, a monitor and a DTMF keypad

A   To operate radioteletype (RTTY) you will need a **modem** and a **teleprinter** or **computer system** along with your transceiver.

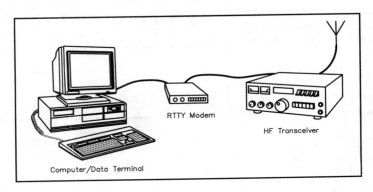

RTTY Modem

HF Transceiver

Computer/Data Terminal

T7A11 In packet-radio operation, what equipment connects to a terminal-node controller?

A. A transceiver and a modem
B. A transceiver and a terminal or computer system
C. A DTMF keypad, a monitor and a transceiver
D. A DTMF microphone, a monitor and a transceiver

B Packet radio uses a special type of modem called a **terminal-node controller**, or **TNC**. Connect a TNC between your **transceiver and your computer system** for packet-radio operation. If you don't have a computer, you use a "dumb **terminal**," which includes a keyboard for data entry and a monitor for data display.

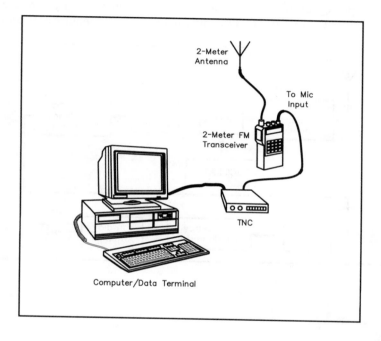

T7B Transmitter and receiver block diagrams; purpose and operation of low-pass, high-pass and band-pass filters

T7B01 What circuit uses a limiter and a frequency discriminator to produce an audio signal?

A. A double-conversion receiver
B. A variable-frequency oscillator
C. A superheterodyne receiver
D. An FM receiver

D   The detector circuit in most **FM receivers** is called a frequency discriminator. The amplitude of the discriminator output changes with variations of the input signal frequency. There is usually a limiter circuit just before the discriminator, to ensure that the discriminator input signal amplitude does not vary.

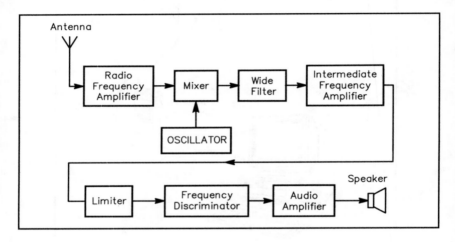

T7B02 What circuit is pictured in Figure T7-1 if block 1 is a variable-frequency oscillator?

   A. A packet-radio transmitter
   B. A crystal-controlled transmitter
   C. A single-sideband transmitter
   D. A VFO-controlled transmitter

   D   A **variable-frequency oscillator (VFO) controls the operating frequency of a transmitter** (or receiver). In a simple CW transmitter, the VFO signal is amplified in the transmitter driver stage, and then the signal is further amplified in a power amplifier. The circuit shown here is a **VFO-controlled transmitter** circuit.

T7B03 What circuit is pictured in Figure T7-1 if block 1 is a crystal oscillator?

   A. A crystal-controlled transmitter
   B. A VFO-controlled transmitter
   C. A single-sideband transmitter
   D. A CW transceiver

   A   When a crystal oscillator controls the operating frequency of a transmitter (or receiver) rather than a variable-frequency oscillator (VFO) the circuit is a **crystal-controlled transmitter**.

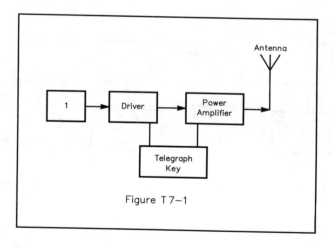

Figure T7-1

T7B04 What type of circuit does Figure T7-2 represent if block 1 is a product detector?

    A. A simple phase modulation receiver
    B. A simple FM receiver
    C. A simple CW and SSB receiver
    D. A double-conversion multiplier

C   This drawing shows a simple **superheterodyne CW and SSB receiver**.

T7B05 If Figure T7-2 is a diagram of a simple single-sideband receiver, what type of circuit should be shown in block 1?

    A. A high pass filter
    B. A ratio detector
    C. A low pass filter
    D. A product detector

D   This drawing shows a simple superheterodyne CW or SSB receiver, and block 1 is a **product detector**.

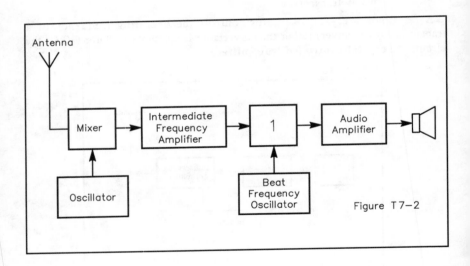

Figure T7-2

T7B06 What circuit is pictured in Figure T7-3, if block 1 is a frequency discriminator?

A. A double-conversion receiver
B. A variable-frequency oscillator
C. A superheterodyne receiver
D. An FM receiver

D   When you look at a circuit block diagram and see a limiter stage and a frequency discriminator, you can be sure it is an **FM receiver**.

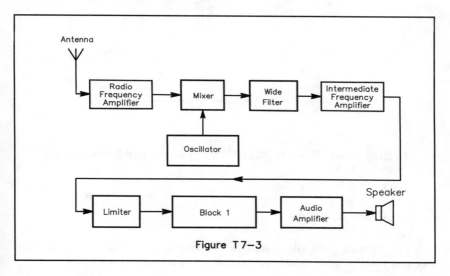

Figure T7-3

T7B07 Why do modern HF transmitters have a built-in low-pass filter in their RF output circuits?

A. To reduce RF energy below a cutoff point
B. To reduce low-frequency interference to other amateurs
C. To reduce harmonic radiation
D. To reduce fundamental radiation

C   A low-pass filter blocks signals above a certain frequency limit and allows those below that limit to pass through. Modern radios include a low-pass filter in the output circuit **to reduce the strength of harmonic signals** before they get to your antenna.

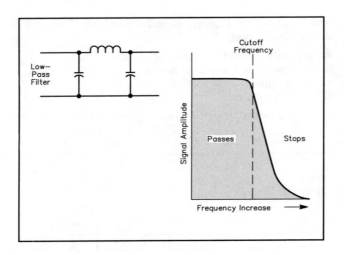

T7B08 What circuit blocks RF energy above and below certain limits?

- A. A band-pass filter
- B. A high-pass filter
- C. An input filter
- D. A low-pass filter

A   A **band-pass filter** blocks RF energy above and below certain frequency limits, while passing signals that fall between those limits.

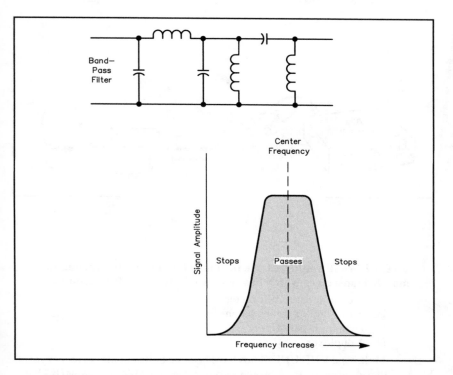

T7B09 What type of filter is used in the IF section of receivers to block energy outside a certain frequency range?

A. A band-pass filter
B. A high-pass filter
C. An input filter
D. A low-pass filter

A See the explanation to question T7B08. One common application for a **band-pass filter** is in the intermediate-frequency (IF) section of a receiver. The filter blocks energy outside a certain frequency range, so only signals inside the desired passband are received.

T7B10 What circuit function is found in all types of receivers?

A. An audio filter
B. A beat-frequency oscillator
C. A detector
D. An RF amplifier

C A receiver **detector** converts the information contained on a modulated radio wave to an audio frequency so you can hear it. Every receiver, from the simplest to the most complex, includes some type of detector circuit.

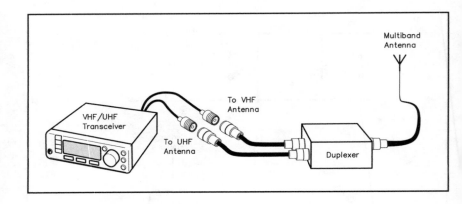

T7B11 What would you use to connect a dual-band antenna to a mobile transceiver which has separate VHF and UHF outputs?

A. A dual-needle SWR meter
B. A full-duplex phone patch
C. Twin high-pass filters
D. A duplexer

D   Multiband VHF/UHF radios have become quite popular in recent years. Most of these radios operate on the popular 2-meter (144 to 148 MHz) band and either the 1.25-meter (222 to 225-MHz) band or the 70-cm (420 to 450-MHz) band. Some of these radios have separate antenna connectors for each band and others have only one connector used for both. You may have one antenna that covers both bands or you may want to use separate antennas for each band. In either case you may need to connect a **duplexer** between your antenna(s) and the radio.

Signals and Emissions

Your Technician exam will include 2 questions from this Signals and Emissions subelement. Those questions will be taken from the 2 groups of questions printed in this chapter.

T8A RF carrier, definition and typical bandwidths; harmonics and unwanted signals; chirp; superimposed hum; equipment and adjustments to help reduce interference to others

T8A01 What is an RF carrier?

A. The part of a transmitter that carries the signal to the transmitter antenna

B. The part of a receiver that carries the signal from the antenna to the detector

C. A radio frequency signal that is modulated to produce a radiotelephone signal

D. A modulation that changes a radio frequency signal to produce a radiotelephone signal

C An RF carrier is a constant-amplitude **radio-frequency (RF) signal**. When the RF carrier is modulated with some type of information, it carries the information signal along with it. If the information signal is a voice then the transmitted signal is called a **radiotelephone signal**.

T8A02 Which list of emission types is in order from the narrowest bandwidth to the widest bandwidth?

A. RTTY, CW, SSB voice, FM voice

B. CW, FM voice, RTTY, SSB voice

C. CW, RTTY, SSB voice, FM voice

D. CW, SSB voice, RTTY, FM voice

C **CW** has the narrowest bandwidth of any emission type commonly used on the amateur bands. **RTTY** has a bandwidth that is a bit wider than a CW signal, and an **SSB voice** signal is wider still. **FM voice** has the widest bandwidth of any voice emission type used by amateurs.

T8A03 What is the usual bandwidth of a single-sideband amateur signal?

   A. 1 kHz
   B. 2 kHz
   C. Between 3 and 6 kHz
   D. Between 2 and 3 kHz

D   A properly adjusted single-sideband voice transmitter produces a signal that has a bandwidth of between **2 kHz** and **3 kHz**.

T8A04 What is the usual bandwidth of a frequency-modulated amateur signal?

   A. Less than 5 kHz
   B. Between 5 and 10 kHz
   C. Between 10 and 20 kHz
   D. Greater than 20 kHz

C   A properly adjusted frequency-modulated (FM) voice transmitter produces a signal that has a bandwidth of between **10 kHz** and **20 kHz**.

T8A05 What is the name for emissions produced by switching a transmitter's output on and off?

   A. Phone
   B. Test
   C. CW
   D. RTTY

C   Morse code, or **CW**, is produced simply by switching the output from a transmitter on and off.

T8A06 What term describes the process of combining an information signal with a radio signal?

   A. Superposition
   B. Modulation
   C. Demodulation
   D. Phase-inversion

B   Every radio signal begins as a steady RF carrier in the transmitter. Then an information signal must be combined with that RF carrier to produce the transmitter signal. The process of combining the RF carrier and the information siganl is called **modulation**.

T8A07 What is the result of over deviation in an FM transmitter?

A. Increased transmitter power
B. Out-of-channel emissions
C. Increased transmitter range
D. Poor carrier suppression

B If the microphone gain is set too high, or the audio signal into the microphone is too loud, an FM transmitter will overdeviate. That means the modulated signal has a wider bandwidth than it should, resulting in signals that are outside of the normal communications channel. Such **out-of-channel emissions** may cause interference to other stations using nearby frequencies.

T8A08 What causes splatter interference?

A. Keying a transmitter too fast
B. Signals from a transmitter's output circuit are being sent back to its input circuit
C. Overmodulation of a transmitter
D. The transmitting antenna is the wrong length

C If the microphone gain is set too high, or the audio signal into the microphone is too loud, an amplitude-modulated (AM) transmitter or single-sideband (SSB) **transmitter will be overmodulated**. That means the transmitted signal has a wider bandwidth than it should, resulting in signals that are outside of the normal communications channel. These signals may cause interference to other stations using nearby frequencies, and this is called **splatter interference**.

T8A09 How does the frequency of a harmonic compare to the desired transmitting frequency?

A. It is slightly more than the desired frequency
B. It is slightly less than the desired frequency
C. It is exactly two, or three, or more times the desired frequency
D. It is much less than the desired frequency

C Harmonics are whole-number multiples of the desired operating frequency. For example, the second harmonic is two times the fundamental frequency, the third harmonic is three times the fundamental, and so on.

T8A10 What should you check if you change your transceiver's microphone from a mobile type to a base station type?

A. Check the CTCSS levels on the oscilloscope
B. Make an on-the-air radio check to ensure the quality of your signal
C. Check the amount of current the transceiver is now using
D. Check to make sure the frequency readout is now correct

B Different microphones can have frequency responses that will affect the way your voice sounds. They can also send different-strength signals to your transmitter, resulting in different modulation characteristics. Any time you change microphones you should make an **on-the-air check** with another station to ensure the quality of your signal. It may be that your voice characteristics and the microphone require a small adjustment to the mike gain or deviation control. This would be especially true if you change microphones from a mobile mike to a base-station-type microphone.

T8A11 Why is good station grounding needed when connecting your computer to your transceiver to receive high-frequency data signals?

A. Good grounding raises the receiver's noise floor
B. Good grounding protects the computer from nearby lightning strikes
C. Good grounding will minimize stray noise on the receiver
D. FCC rules require all equipment to be grounded

C More and more amateurs are using computers in their ham shacks. Your computer can cause interference to your receiver because the computer has a clock oscillator that operates at a *radio* frequency. For example, with a Pentium or 433 or 866 computer, the 433 and 866 represent the clock frequency in megahertz! The computer should be in a metal cabinet, with all the screws securely attached, and you should use shielded cables with the shield connected to the equipment chassis. **Be sure all equipment is properly grounded.** These steps will help reduce the possibility of interference.

T8B Concepts and types of modulation: CW, phone, RTTY and data emission types; FM deviation

T8B01 What is the name for packet-radio emissions?

A. CW
B. Data
C. Phone
D. RTTY

B Packet-radio signals are a form of communication designed to operate between two computer systems, often automatically. The communication uses computer codes to exchange files and information, so it is called a **data** emission.

T8B02 What is the name of the voice emission most used on VHF/UHF repeaters?

A. Single-sideband phone
B. Pulse-modulated phone
C. Slow-scan phone
D. Frequency-modulated phone

D   Most VHF and UHF repeater operators use **frequency-modulated** (FM) **phone** for voice communications.

T8B03 What is meant by the upper-sideband (USB)?

A. The part of a single-sideband signal that is above the carrier frequency
B. The part of a single-sideband signal that is below the carrier frequency
C. Any frequency above 10 MHz
D. The carrier frequency of a single-sideband signal

A   When an RF carrier signal is amplitude modulated by an audio signal, such as a voice, two new signals are produced. One of these comes from adding the carrier frequency and the audio frequency. This new signal has a frequency that is higher than the carrier, so it is called the **upper sideband** signal. The other signal is produced by subtracting the audio frequency from the carrier frequency, and it has a frequency that is lower than the carrier. This is called the **lower sideband**. Either sideband may be selected for transmission in a single-sideband (SSB) transmitter.

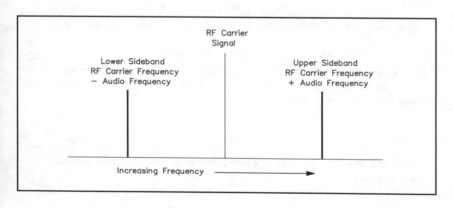

T8B04 What does the term "phone transmissions" usually mean?

    A. The use of telephones to set up an amateur contact

    B. A phone patch between amateur radio and the telephone system

    C. AM, FM or SSB voice transmissions by radiotelephony

    D. Placing the telephone handset near a transceiver's microphone and speaker to relay a telephone call

C   When your transmitter sends any voice communication mode, it is called *phone* operation under FCC Rules. **AM, FM and SSB voice are all types of phone emission.**

T8B05 How is an HF RTTY signal usually produced?

    A. By frequency-shift keying an RF signal

    B. By on/off keying an RF signal

    C. By digital pulse-code keying of an unmodulated carrier

    D. By on/off keying an audio-frequency signal

A   On HF, radioteletype (RTTY) is usually transmitted by shifting the frequency of the RF signal between two frequencies — the mark and space frequencies. This technique is called **frequency-shift keying** of the RF signal.

T8B06 What are two advantages to using modern data-transmission techniques for communications?

    A. Very simple and low-cost equipment

    B. No parity-checking required and high transmission speed

    C. Easy for mobile stations to use and no additional cabling required

    D. High transmission speed and communications reliability

D   Two popular forms of data transmission are radioteletype and packet radio. These modes provide **high transmission speeds and good communications reliability**.

T8B07 Which sideband is commonly used for 10-meter phone operation?

    A. Upper sideband

    B. Lower sideband

    C. Amplitude-compandored sideband

    D. Double sideband

A   Either upper or lower sideband can technically be used for phone operation on any band. Amateurs normally use **upper sideband** for phone operation on the 10-meter band.

T8B08 What can you do if you are told your FM hand-held or mobile transceiver is over-deviating?

A. Talk louder into the microphone
B. Let the transceiver cool off
C. Change to a higher power level
D. Talk farther away from the microphone

D   If you are told the transmitted audio from your FM transceiver is distorted or over-deviating, you should **try talking farther away from the microphone** and speaking with a quieter voice.

T8B09 What does chirp mean?

A. An overload in a receiver's audio circuit whenever CW is received
B. A high-pitched tone that is received along with a CW signal
C. A small change in a transmitter's frequency each time it is keyed
D. A slow change in transmitter frequency as the circuit warms up

C   If the transmitter power supply voltage changes when more current is drawn from the supply, the transmitter operating frequency may change each time the transmitter is keyed. This **small change in transmitter frequency is called chirp.**

Antennas and Feed Lines

There will be 2 questions from the Antennas and Feed Lines subelement on your Technician exam. Those questions will be taken from the 2 groups of questions printed in this chapter.

T9A   Wavelength vs. antenna length; 1/2 wavelength dipole and 1/4 wavelength vertical antennas; multiband antennas

T9A01 How do you calculate the length (in feet) of a half-wavelength dipole antenna?

A. Divide 150 by the antenna's operating frequency (in MHz) [150/f (in MHz)]

B. Divide 234 by the antenna's operating frequency (in MHz) [234/f (in MHz)]

C. Divide 300 by the antenna's operating frequency (in MHz) [300/f (in MHz)]

D. Divide 468 by the antenna's operating frequency (in MHz) [468/f (in MHz)]

D   Calculate the length of a half-wavelength dipole antenna for use on the HF bands by **dividing 468 by the desired operating frequency in megahertz (MHz)**.

$$\text{Length (in feet) for 1/2-}\lambda \text{ dipole antenna} = \frac{468}{f\,(\text{in MHz})}$$

This equation includes a factor to account for something called *end effects* and other antenna factors, and is generally not as accurate for VHF and UHF antennas.

T9A02 How do you calculate the length (in feet) of a quarter-wavelength vertical antenna?

    A. Divide 150 by the antenna's operating frequency (in MHz) [150/f (in MHz)]

    B. Divide 234 by the antenna's operating frequency (in MHz) [234/f (in MHz)]

    C. Divide 300 by the antenna's operating frequency (in MHz) [300/f (in MHz)]

    D. Divide 468 by the antenna's operating frequency (in MHz) [468/f (in MHz)]

B Calculate the length of a quarter-wavelength vertical antenna for use on the HF bands by **dividing 234 by the desired operating frequency in megahertz (MHz).**

$$\text{Length (in feet) for 1/4-}\lambda \text{ vertical antenna} = \frac{234}{f\,(\text{in MHz})}$$

This equation includes a factor to account for something called *end effects* and other antenna factors, and is generally not as accurate for VHF and UHF antennas.

T9A03 How long should you make a quarter-wavelength vertical antenna for 440 MHz (measured to the nearest inch)?

    A. 12 inches

    B. 9 inches

    C. 6 inches

    D. 3 inches

C Calculate the length of a quarter-wavelength vertical antenna for use on the HF bands by **dividing 234 by the desired operating frequency in megahertz (MHz).**

$$\text{Length (in feet) for 1/4-}\lambda \text{ vertical antenna} = \frac{234}{f\,(\text{in MHz})}$$

Although this equation is not as accurate for VHF and UHF antennas, it can provide a pretty good estimate for the length of quarter-wave vertical antennas for these bands, too. The answer, 0.53, about half a foot, can be rounded off to **6 inches**.

T9A04 How long should you make a quarter-wavelength vertical antenna for 28.450 MHz (measured to the nearest foot)?

A. 8 ft
B. 12 ft
C. 16 ft
D. 24 ft

A   Calculate the length of a quarter-wavelength vertical antenna for use on the HF bands by **dividing 234 by the desired operating frequency in megahertz (MHz)**.

$$\text{Length (in feet) for 1/4-}\lambda \text{ vertical antenna} = \frac{234}{f\,(\text{in MHz})}$$

This equation includes a factor to account for something called *end effects* and other antenna factors.

The answer is 8.22, about **8 feet**.

T9A05 How long should you make a quarter-wavelength vertical antenna for 146 MHz (measured to the nearest inch)?

A. 112 inches
B. 50 inches
C. 19 inches
D. 12 inches

C   Calculate the length of a quarter-wavelength vertical antenna for use on the HF bands by **dividing 234 by the desired operating frequency in megahertz (MHz)**.

$$\text{Length (in feet) for 1/4-}\lambda \text{ vertical antenna} = \frac{234}{f\,(\text{in MHz})}$$

Although this equation is not as accurate for VHF and UHF antennas, it can provide a pretty good estimate for the length of quarter-wave vertical antennas for these bands, too. The answer, 1.6 feet, can be converted to inches by multiplying by 12. The answer in inches is 19.2, which can be rounded to **19 inches**.

T9A06 If an antenna is made longer, what happens to its resonant frequency?

A. It decreases
B. It increases
C. It stays the same
D. It disappears

A   Antenna lengths are related to the wavelength of the desired signal. As the wavelength of a radio signal gets longer the frequency decreases. Therefore, if an antenna is made longer it will have a lower resonant frequency than the shorter antenna. **The resonant frequency of an antenna decreases as the antenna is made longer.**

T9A07 If an antenna is made shorter, what happens to its resonant frequency?

A. It decreases
B. It increases
C. It stays the same
D. It disappears

B   Antenna lengths are related to the wavelength of the desired signal. As the wavelength of a radio signal gets shorter the frequency increases. Therefore, if an antenna is made shorter it will have a higher resonant frequency than the longer antenna. **The resonant frequency of an antenna increases as the antenna is made shorter.**

T9A08 How could you decrease the resonant frequency of a dipole antenna?

A. Lengthen the antenna
B. Shorten the antenna
C. Use less feed line
D. Use a smaller size feed line

A   Antenna lengths are related to the wavelength of the desired signal. As the wavelength of a radio signal gets longer the frequency gets lower. Therefore, **if you lengthen an antenna it will have a lower resonant frequency** than the shorter antenna.

T9A09 How could you increase the resonant frequency of a dipole antenna?

A. Lengthen the antenna
B. Shorten the antenna
C. Use more feed line
D. Use a larger size feed line

B Antenna lengths are related to the wavelength of the desired signal. As the wavelength of a radio signal gets shorter the frequency increases. You can increase the resonant frequency of an antenna by **making the antenna shorter**.

T9A10 What is one advantage to using a multiband antenna?

A. You can operate on several bands with a single feed line
B. Multiband antennas always have high gain
C. You can transmit on several frequencies simultaneously
D. Multiband antennas offer poor harmonic suppression

A Almost any type of antenna can be made into a multiband antenna. Dipole antennas, vertical antennas and various types of beam antennas can be designed for operation on several bands. One main advantage to any multiband antenna is that **you can operate on several bands with one antenna and a single feed line**.

T9A11 What is one disadvantage to using a multiband antenna?

A. It must always be used with a balun
B. It will always have low gain
C. It cannot handle high power
D. It can radiate unwanted harmonics

D A multiband antenna will radiate signals on two or more bands. A poorly adjusted transmitter, or a transmitter without the proper low-pass filter on the output may produce harmonic signals. If those harmonic signals fall on one of the bands for which the antenna is designed, it will **radiate those unwanted harmonics** efficiently.

T9B Parasitic beam directional antennas; polarization, impedance matching and SWR, feed lines, balanced vs. unbalanced (including baluns)

T9B01 What is a directional antenna?

A. An antenna that sends and receives radio energy equally well in all directions

B. An antenna that cannot send and receive radio energy by skywave or skip propagation

C. An antenna that sends and receives radio energy mainly in one direction

D. An antenna that uses a directional coupler to measure power transmitted

C A *beam*, or directional antenna, **concentrates the radio energy into a particular direction**. The transmitted signal is stronger in the *forward*, or desired direction and weaker in other directions. It is important to realize that a beam antenna doesn't "create" more power. It just concentrates more of the transmitter power in the forward direction. A beam antenna also receives signals more effectively from the forward direction and rejects signals from other directions.

T9B02 How is a Yagi antenna constructed?

A. Two or more straight, parallel elements are fixed in line with each other

B. Two or more square or circular loops are fixed in line with each other

C. Two or more square or circular loops are stacked inside each other

D. A straight element is fixed in the center of three or more elements that angle toward the ground

A A Yagi type of beam antenna is made using two or more elements that are placed parallel to each other. The *driven element* is approximately 1/2 wavelength long, and the feed line connects to the center of this element. There may be one *reflector element*, which is slightly longer, and is placed behind the driven element. There may be one or more *director elements*, which are slightly shorter, and placed in front of the driven element.

T9B03 How many directly driven elements do most parasitic beam antennas have?

A. None

B. One

C. Two

D. Three

B Parasitic beam antennas usually have **one** directly driven element.

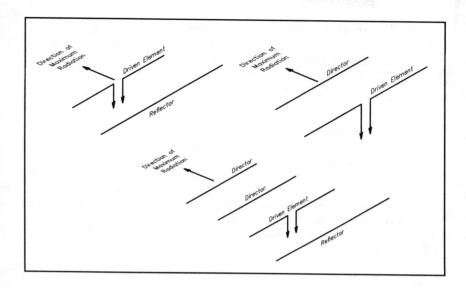

T9B04 What is a parasitic beam antenna?

A. An antenna in which some elements obtain their radio energy by induction or radiation from a driven element

B. An antenna in which wave traps are used to magnetically couple the elements

C. An antenna in which all elements are driven by direct connection to the feed line

D. An antenna in which the driven element obtains its radio energy by induction or radiation from director elements

A   A parasitic beam antenna has one directly driven element and one or more elements that **receive radio energy by induction or radiation from the driven element**. A Yagi antenna is one example of a parasitic beam antenna.

T9B05 What are the parasitic elements of a Yagi antenna?

A. The driven element and any reflectors

B. The director and the driven element

C. Only the reflectors (if any)

D. Any directors or any reflectors

D   See the explanation for question T9B02. The parasitic elements of a Yagi beam antenna are **any directors and the reflector**.

T9B06 **What is a cubical quad antenna?**

    A. Four straight, parallel elements in line with each other, each approximately 1/2-electrical wavelength long

    B. Two or more parallel four-sided wire loops, each approximately one-electrical wavelength long

    C. A vertical conductor 1/4-electrical wavelength high, fed at the bottom

    D. A center-fed wire 1/2-electrical wavelength long

B   A cubical quad antenna is made using **two or more four-sided wire loops** placed parallel to each other. (*Quad* means four, and *cubical* tells us the four sides have equal lengths.) The *driven element* is approximately **1 wavelength long**, and the feed line connects to this element. There may be one *reflector element*, which is made from a loop of wire that is slightly longer, and is placed behind the driven element. There may be one or more *director elements*, which are made from a loop of wire that is slightly shorter, and placed in front of the driven element.

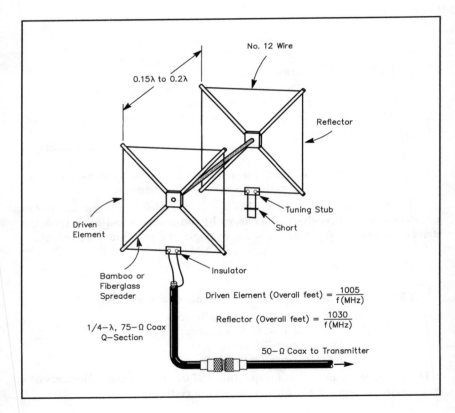

No. 12 Wire

0.15λ to 0.2λ

Reflector

Tuning Stub

Short

Driven Element

Insulator

Bamboo or Fiberglass Spreader

$$\text{Driven Element (Overall feet)} = \frac{1005}{f\,(\text{MHz})}$$

$$\text{Reflector (Overall feet)} = \frac{1030}{f\,(\text{MHz})}$$

1/4–λ, 75–Ω Coax Q–Section

50–Ω Coax to Transmitter

T9B07 What type of non-directional antenna is easy to make at home and works well outdoors?

A. A Yagi
B. A delta loop
C. A cubical quad
D. A ground plane

D   A **ground-plane** antenna is a type of non-directional antenna (it radiates signals equally well in all compass directions) that is easy to build and gives good performance.

T9B08 What electromagnetic-wave polarization does most man-made electrical noise have in the HF and VHF spectrum?

A. Horizontal
B. Left-hand circular
C. Right-hand circular
D. Vertical

D   Most man-made electrical noise has a **vertical** wave polarization.

T9B09 What does standing-wave ratio mean?

A. The ratio of maximum to minimum inductances on a feed line
B. The ratio of maximum to minimum capacitances on a feed line
C. The ratio of maximum to minimum impedances on a feed line
D. The ratio of maximum to minimum voltages on a feed line

D   The standing-wave ratio (SWR) on a feed line is equal to the **ratio of the maximum voltage** along the line **to the minimum voltage** along the line. SWR gives an indication of how well the feed line impedance matches the antenna input impedance. An SWR of 1:1 indicates a perfect match while higher values, such as 3:1, indicate an impedance mismatch.

T9B10 Where would you install a balun to feed a dipole antenna with 50-ohm coaxial cable?

A. Between the coaxial cable and the antenna
B. Between the transmitter and the coaxial cable
C. Between the antenna and the ground
D. Between the coaxial cable and the ground

A   Install a balun directly at the antenna, **between the coaxial cable and the antenna.**

T9B11 Why does coaxial cable make a good antenna feed line?

- A. You can make it at home, and its impedance matches most amateur antennas
- B. It is weatherproof, and it can be used near metal objects
- C. It is weatherproof, and its impedance is higher than that of most amateur antennas
- D. It can be used near metal objects, and its impedance is higher than that of most amateur antennas

B The most common types of coaxial cable have a characteristic impedance of about 50 or 75 ohms, which provides a good impedance match to most amateur transmitters and many common antenna types. Coaxial cable is **weatherproof**, and it is generally **unaffected by nearby metal objects**.

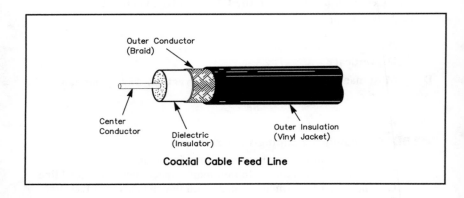

Coaxial Cable Feed Line

T0

RF Safety

Your Technician exam will include 3 questions from the RF Safety subelement, and those questions will be taken from the 3 groups of questions designated as T0A, T0B and T0C.

The questions in this chapter represent the VEC Question Pool Committee's efforts to present you with questions that meet the FCC requirements and also help you learn about this important topic. These questions will help you learn how to evaluate your amateur station to ensure that it doesn't exceed the FCC maximum permissible exposure (MPE) limits for human exposure to RF radiation.

The short answers to these questions have been adapted from text that was written and approved by the members of the ARRL RF Safety Committee. We recommend that you read the entire RF Safety text, as printed in the fourth edition of ARRL's *Now You're Talking!* or the latest editions of *The ARRL Handbook, The ARRL Antenna Book* or *The ARRL Operating Manual.* If you find any of these questions or their answers to be confusing or unclear, the additional material will give you a better understanding.

T0A RF safety fundamentals, terms and definitions

T0A01 Why is it a good idea to adhere to the FCC's Rules for using the minimum power needed when you are transmitting with your hand-held radio?

A. Large fines are always imposed on operators violating this rule
B. To reduce the level of RF radiation exposure to the operator's head
C. To reduce calcification of the NiCd battery pack
D. To eliminate self oscillation in the receiver RF amplifier

B   With hand-held transceivers, keep the antenna away from your head and use the lowest power possible to maintain communications. Use a separate microphone and hold the rig as far away from you as possible. This will **reduce your exposure to the RF energy**.

T0A02 Over what frequency range are the FCC Regulations most stringent for RF radiation exposure?

A. Frequencies below 300 kHz
B. Frequencies between 300 kHz and 3 MHz
C. Frequencies between 3 MHz and 30 MHz
D. Frequencies between 30 MHz and 300 MHz

D   At frequencies near the body's natural resonant frequency, RF energy is absorbed more efficiently, and maximum heating occurs. In adults, this frequency usually is about 35 MHz if the person is grounded, and about 70 MHz if the person's body is insulated from the ground. Also, body parts may be resonant; the adult head, for example is resonant around 400 MHz, while a baby's smaller head resonates near 700 MHz. Body size thus determines the frequency at which most RF energy is absorbed. As the frequency is increased above resonance, less absorption generally occurs. However, additional longitudinal resonances occur at about 1 GHz near the body surface. *Specific absorption rate (SAR)* is a term that describes the rate at which RF energy is absorbed into the human body. Maximum permissible exposure (MPE) limits are based on whole-body SAR values. This helps explain why these safe exposure limits vary with frequency.

The **lowest (most stringent)** maximum permissible exposure **(MPE) limits** over the range of frequencies covered by the FCC Rules are from **30 to 300 MHz.**

T0A03 What is one biological effect to the eye that can result from RF exposure?

A. The strong magnetic fields can cause blurred vision
B. The strong magnetic fields can cause polarization lens
C. It can cause heating, which can result in the formation of cataracts
D. It can cause heating, which can result in astigmatism

C   Body tissues that are subjected to very high levels of RF energy may suffer serious **heat damage**. These effects depend upon the frequency of the energy, the power density of the RF field that strikes the body, and even on factors such as the polarization of the wave. In extreme cases, **RF-induced heating in the eye can result in cataract formation** and can even cause blindness. Excessive RF heating of the reproductive organs can cause sterility. Other serious health problems can also result from RF heating. These heat-related health hazards are called *thermal effects.*

T0A04 In the far field, as the distance from the source increases, how does power density vary?

A. The power density is proportional to the square of the distance
B. The power density is proportional to the square root of the distance
C. The power density is proportional to the inverse square of the distance
D. The power density is proportional to the inverse cube of the distance

C   The radiating far-field forms the traveling electromagnetic waves. One distinguishing characteristic of far-field radiation is that the **power density is proportional to the inverse square of the distance.** (That means if you double the distance from the antenna, the power density will be one fourth as strong.)

T0A05 In the near field, how does the field strength vary with distance from the source?

A. It always increases with the cube of the distance
B. It always decreases with the cube of the distance
C. It varies as a sine wave with distance
D. It depends on the type of antenna being used

D   The strength of the reactive near field decreases in a complicated fashion as you increase the distance from the antenna. Beyond the reactive near field, the antenna's radiated field is divided into two other regions: the *radiating near field* and the *radiating far field*. Nearly any metal object or other conductor that is located within the radiating near field can alter the radiation pattern of the antenna. Conductors such as telephone wiring or aluminum siding on a building will interact with the theoretical electric and magnetic fields to add or subtract intensity. This results in areas of varying field strength. Although you have measured the fields in the general area around your antenna and found that your station meets the MPE limits, there may still be "hot spots" or areas of higher field strengths within that region. **In the near field of an antenna, the field strength varies in a way that depends on the type of antenna and other nearby objects as you move farther away from the antenna.**

TOA06 Why should you never look into the open end of a microwave feed horn antenna while the transmitter is operating?

A. You may be exposing your eyes to more than the maximum permissible exposure of RF radiation
B. You may be exposing your eyes to more than the maximum permissible exposure level of infrared radiation
C. You may be exposing your eyes to more than the maximum permissible exposure level of ultraviolet radiation
D. All of these choices are correct

A   In the UHF/SHF region, never look into the open end of an activated length of waveguide or microwave feed-horn antenna or point it toward anyone. (If you do, **you may be exposing your eyes to more than the maximum permissible exposure level of RF radiation.**)

TOA07 Why are Amateur Radio operators required to meet the FCC RF radiation exposure limits?

A. The standards are applied equally to all radio services
B. To ensure that RF radiation occurs only in a desired direction
C. Because amateur station operations are more easily adjusted than those of commercial radio services
D. To ensure a safe operating environment for amateurs, their families and neighbors

D   The FCC regulations control *exposure* to RF fields, not the *strength* of RF fields. There is no limit to how strong a field can be as long as no one is being exposed to it, although FCC regulations require that amateurs use the minimum necessary power at all times (§97.313 [a]). **All radio stations must comply with the requirements for MPE limits**, even QRP stations running only a few watts or less. The MPEs vary with frequency. The FCC requires that certain amateur stations be evaluated for compliance with the MPEs. This will **help ensure a safe operating environment for amateurs, their families and neighbors**.

TOA08 Why are the maximum permissible exposure (MPE) levels not uniform throughout the radio spectrum?

A. The human body absorbs energy differently at various frequencies

B. Some frequency ranges have a cooling effect while others have a heating effect on the body

C. Some frequency ranges have no effect on the body

D. Radiation at some frequencies can have a catalytic effect on the body

A   At frequencies near the body's natural resonant frequency, RF energy is absorbed more efficiently, and maximum heating occurs. In adults, this frequency usually is about 35 MHz if the person is grounded, and about 70 MHz if the person's body is insulated from the ground. Also, body parts may be resonant; the adult head, for example is resonant around 400 MHz, while a baby's smaller head resonates near 700 MHz. Body size thus determines the frequency at which most RF energy is absorbed. As the frequency is increased above resonance, less RF heating generally occurs. However, additional longitudinal resonances occur at about 1 GHz near the body surface. *Specific absorption rate (SAR)* is a term that describes the rate at which RF energy is absorbed into the human body. Maximum permissible exposure (MPE) limits are based on whole-body SAR values. This helps explain why these safe exposure limits vary with frequency.

The lowest (most stringent) maximum permissible exposure (MPE) limits over the range of frequencies covered by the FCC Rules are from 30 to 300 MHz.

TOA09 What does the term "specific absorption rate" or SAR mean?

A. The degree of RF energy consumed by the ionosphere

B. The rate at which transmitter energy is lost because of a poor feed line

C. The rate at which RF energy is absorbed into the human body

D. The amount of signal weakening caused by atmospheric phenomena

C   At frequencies near the body's natural resonant frequency, RF energy is absorbed more efficiently, and maximum heating occurs. In adults, this frequency usually is about 35 MHz if the person is grounded, and about 70 MHz if the person's body is insulated from the ground. Also, body parts may be resonant; the adult head, for example is resonant around 400 MHz, while a baby's smaller head resonates near 700 MHz. Body size thus determines the frequency at which most RF energy is absorbed. As the frequency is increased above resonance, less RF heating generally occurs. However, additional longitudinal resonances occur at about 1 GHz near the body surface. **Specific absorption rate (SAR) is** a term that describes **the rate at which RF energy is absorbed into the human body**. Maximum permissible exposure (MPE) limits are based on whole-body SAR values. This helps explain why these safe exposure limits vary with frequency.

TOA10 On what value are the maximum permissible exposure (MPE) limits based?

A. The square of the mass of the exposed body
B. The square root of the mass of the exposed body
C. The whole-body specific gravity (WBSG)
D. The whole-body specific absorption rate (SAR)

D Specific absorption rate (SAR) is a term that describes the rate at which RF energy is absorbed into the human body. **Maximum permissible exposure (MPE) limits are based on whole-body SAR values.** These safe exposure limits vary with frequency because at frequencies near the body's natural resonant frequency, RF energy is absorbed more efficiently, and maximum heating occurs.

T0B RF safety rules and guidelines

T0B01 Where will you find the applicable FCC RF radiation maximum permissible exposure (MPE) limits defined?

A. FCC Part 97 Amateur Service Rules and Regulations
B. FCC Part 15 Radiation Exposure Rules and Regulations
C. FCC Part 1 and Office of Engineering and Technology (OET) Bulletin 65
D. Environmental Protection Agency Regulation 65

C Amateurs normally look only to Part 97 for the Rules governing Amateur Radio. There are other Parts to the FCC Rules, though. **Part 1**, for example, contains the exact limits for exposure to RF radiation.

Part 97 also refers to RF exposure limits, however. § 97.13 (c) (2) states [emphasis added]:

> If the routine environmental evaluation indicates that the RF electromagnetic fields could exceed the limits contained in § 1.1310 of this chapter in accessible areas, the licensee must take action to prevent human exposure to such RF electromagnetic fields. **Further information** on evaluating compliance with these limits **can be found in the OET Bulletin Number 65**, "Evaluating Compliance with FCC Guidelines for Human Exposure to Radiofrequency Electromagnetic Fields."

T0B02 What factors must you consider if your repeater station antenna will be located at a site that is occupied by antennas for transmitters in other services?

A. Your radiated signal must be considered as part of the total RF radiation from the site when determining RF radiation exposure levels

B. Each individual transmitting station at a multiple transmitter site must meet the RF radiation exposure levels

C. Each station at a multiple-transmitter site may add no more than 1% of the maximum permissible exposure (MPE) for that site

D. Amateur stations are categorically excluded from RF radiation exposure evaluation at multiple-transmitter sites

A   If you are installing a repeater or other transmitter in a location that includes antennas and transmitters operating in other services, you must be aware that **the total site installation must meet the FCC RF radiation MPE limits**. This means **your signal is only one part of the total RF radiation from that location**. You will probably have to cooperate with the licensees for the other transmitters to determine the total exposure.

T0B03 To determine compliance with the maximum permitted exposure (MPE) levels, safe exposure levels for RF energy are averaged for an "uncontrolled" RF environment over what time period?

A. 6 minutes
B. 10 minutes
C. 15 minutes
D. 30 minutes

D   One step you can take to limit your RF radiation exposure is by reducing your actual transmitting time. The FCC regulations specify time averaged MPE limits. For an **uncontrolled RF exposure environment**, the exposure is **averaged over any 30-minute period**. So if your routine RF radiation exposure evaluation indicates that you might exceed the MPE limits for an uncontrolled RF environment, reduce your actual transmit time during any 30-minute period.

TOB04 To determine compliance with the maximum permitted exposure (MPE) levels, safe exposure levels for RF energy are averaged for a "controlled" RF environment over what time period?

A. 6 minutes
B. 10 minutes
C. 15 minutes
D. 30 minutes

A   One step you can take to limit your RF radiation exposure is by reducing your actual transmitting time. The FCC regulations specify time averaged MPE limits. For a **controlled RF exposure environment**, the exposure is **averaged over any 6-minute period**. If your routine RF radiation exposure evaluation indicates that you might exceed the MPE limits for a controlled RF environment, reduce your actual transmit time during any 6-minute period.

TOB05 Which of the following categories describes most common amateur use of a hand-held transceiver?

A. Mobile devices
B. Portable devices
C. Fixed devices
D. None of these choices is correct

B   Hand-held radios are very popular for VHF and UHF operation, especially with FM repeaters. They transmit with less than 7 watts of power, which is generally considered safe. Because the radios are designed to be operated with an antenna that is within 20 centimeters of your body, they are classified as **portable devices** by the FCC. Some special considerations are in order to ensure safe operation. This is especially true because hand-held radios generally place the antenna close to your head. Try to position the radio so the antenna is as far from your head (and especially your eyes) while transmitting. An external speaker microphone can be helpful.

TOB06 How does an Amateur Radio operator demonstrate that he or she has read and understood the FCC rules about RF-radiation exposure?

A. By indicating his or her understanding of this requirement on an amateur license application form at the time of application
B. By posting a copy of Part 97 at the station
C. By completing an FCC Environmental Assessment Form
D. By completing an FCC Environmental Impact Statement

A   The FCC will ask you to demonstrate that you have read and understood the FCC Rules about RF-radiation exposure **by indicating that understanding on** FCC Form 605, **the application form for a license, when you apply for one.**

TOB07 What amateur stations must comply with the requirements for RF radiation exposure spelled out in Part 97?

A. Stations with antennas that exceed 10 dBi of gain.
B. Stations that have a duty cycle greater than 50 percent.
C. Stations that run more than 50 watts peak envelope power (PEP)
D. All amateur stations regardless of power

D **All radio stations must comply** with the requirements for maximum permissible exposure (MPE) limits, even QRP (low power) stations running only a few watts or less. Some types of amateur stations do not need to be evaluated to determine their RF radiation exposure, but these stations must still comply with the MPE limits. The station licensee remains responsible for ensuring that the station meets these requirements.

TOB08 Who is responsible for ensuring that an amateur station complies with FCC Rules about RF radiation exposure?

A. The Federal Communications Commission
B. The Environmental Protection Agency
C. The licensee of the amateur station
D. The Food and Drug Administration

C **The station licensee is responsible** for ensuring that the station meets the RF radiation maximum permissible exposure (MPE) requirements.

TOB09 Why do exposure limits vary with frequency?

A. Lower-frequency RF fields have more energy than higher-frequency fields
B. Lower-frequency RF fields penetrate deeper into the body than higher-frequency fields
C. The body's ability to absorb RF energy varies with frequency
D. It is impossible to measure specific absorption rates at some frequencies

C **RF energy is absorbed more efficiently at frequencies near the body's natural resonant frequency, and maximum heating occurs.** In adults, this frequency usually is about 35 MHz if the person is grounded, and about 70 MHz if the person's body is insulated from the ground. Also, body parts may be resonant; the adult head, for example is resonant around 400 MHz, while a baby's smaller head resonates near 700 MHz. Body size thus determines the frequency at which most RF energy is absorbed. As the frequency is increased above resonance, less RF heating generally occurs. However, additional longitudinal resonances occur at about 1 GHz near the body surface. Specific absorption rate (SAR) is a term that describes the rate at which RF energy is absorbed into the human body. **Maximum permissible exposure (MPE) limits are based on whole-body SAR values.** This helps explain why these safe exposure limits vary with frequency.

TOB10 Why is the concept of "duty cycle" one factor used to determine safe RF radiation exposure levels?

    A. It takes into account the amount of time the transmitter is operating at full power during a single transmission
    B. It takes into account the transmitter power supply rating
    C. It takes into account the antenna feed line loss
    D. It takes into account the thermal effects of the final amplifier

    A   Everything else being equal, some emission modes will result in less RF radiation exposure than others. For example, modes like RTTY or FM voice transmit at full power during the entire transmission. On CW, you transmit at full power during dots and dashes and at zero power during the space between these elements. A single-sideband (SSB) phone signal generally produces a lower radiation exposure than a CW or FM transmitter because the transmitter is at full power for only a small percentage of the time during a single transmission. The duty cycle of an emission is the ratio of the average power to the peak envelope power (PEP) of a transmission, expressed as a percentage. In effect, **the concept of duty cycle takes into account the time a transmitter is operating at full power during a single transmission.** An emission with a lower duty cycle produces less RF radiation exposure for the same PEP output.

TOB11 From an RF safety standpoint, what impact does the duty cycle have on the minimum safe distance separating an antenna and the neighboring environment?

    A. The lower the duty cycle, the shorter the compliance distance
    B. The compliance distance is increased with an increase in the duty cycle
    C. Lower duty cycles subject the environment to lower radio-frequency radiation cycles
    D. All of these answers are correct

    D   Everything else being equal, some emission modes will result in less RF radiation exposure than others. For example, modes like RTTY or FM voice transmit at full power during the entire transmission. On CW, you transmit at full power during dots and dashes and at zero power during the space between these elements. A single-sideband (SSB) phone signal generally produces a lower radiation exposure than a CW or FM transmitter because the transmitter is at full power for only a small percentage of the time during a single transmission. The duty cycle of an emission is the ratio of the average power to the peak envelope power (PEP) of a transmission, expressed as a percentage. In effect, the concept of duty cycle takes into account the time a transmitter is operating at full power during a single transmission. **An emission with a lower duty cycle produces less RF radiation exposure for the same PEP output.**

    Lower duty cycles, then, **result in lower RF radiation exposures**. That also means the **antenna can be closer to people** without exceeding their MPE limits. Compared to a 100% duty-cycle mode, people can be closer to your antenna if you are using a 50% duty-cycle mode.**All of the answer choices to this question are correct.**

T0C Routine station evaluation (Practical applications for VHF/UHF and above operations)

T0C01 If you do not have the equipment to measure the RF power densities present at your station, what might you do to ensure compliance with the FCC RF radiation exposure limits?

A. Use one or more of the methods included in the amateur supplement to FCC OET Bulletin 65

B. Call an FCC-Certified Test Technician to perform the measurements for you

C. Reduce power from 200 watts PEP to 100 watts PEP

D. Operate only low-duty-cycle modes such as FM

A   The FCC requires that certain amateur stations be evaluated for compliance with the maximum permissible exposure (MPE) limits. This will help ensure a safe operating environment for amateurs, their families and neighbors. Although an amateur can have someone else do the evaluation, it is not difficult for hams to evaluate their own stations. **FCC Bulletin 65** and the **Amateur Supplement** to that Bulletin **contains basic information** about the regulations and a number of tables that show compliance distances for specific antennas and power levels. Generally, hams will **use these tables to evaluate their stations**. If they choose, however, they can do more extensive calculations, use a computer to model their antenna and exposure, or make actual measurements.

T0C02 Is it necessary for you to perform mathematical calculations of the RF radiation exposure if your station transmits with more than 50 watts peak envelope power (PEP)?

A. Yes, calculations are always required to ensure greatest accuracy

B. Calculations are required if your station is located in a densely populated neighborhood

C. No, calculations may not give accurate results, so measurements are always required

D. No, there are alternate means to determine if your station meets the RF radiation exposure limits

D   The FCC requires that certain amateur stations be evaluated for compliance with the maximum permissible exposure (MPE) limits. This will help ensure a safe operating environment for amateurs, their families and neighbors. Although an amateur can have someone else do the evaluation, it is not difficult for hams to evaluate their own stations. FCC Bulletin 65 and the Amateur Supplement to that Bulletin contains basic information about the regulations and a number of tables that show compliance distances for specific antennas and power levels. Generally, hams will use these tables to evaluate their stations. If they choose, however, they can do more extensive calculations, use a computer to model their antenna and exposure, or make actual measurements.

The important thing to keep in mind is that **there are a variety of possible methods for evaluating the radiation exposure** from your station. You don't have to perform mathematical calculations if you want to choose one of the alternatives.

TOC03 Why should you make sure the antenna of a hand-held transceiver is not too close to your head when transmitting?

    A. To help the antenna radiate energy equally in all directions
    B. To reduce your exposure to the radio-frequency energy
    C. To use your body to reflect the signal in one direction
    D. To keep electrostatic charges from harming the operator

B   With hand-held transceivers, keep the antenna away from your head and use the lowest power possible to maintain communications. If possible, use a separate microphone and hold the rig as far away from you as possible. This will **reduce your exposure to the RF energy**.

TOC04 What should you do for safety if you put up a UHF transmitting antenna?

    A. Make sure the antenna will be in a place where no one can get near it when you are transmitting
    B. Make sure that RF field screens are in place
    C. Make sure the antenna is near the ground to keep its RF energy pointing in the correct direction
    D. Make sure you connect an RF leakage filter at the antenna feed point

A   In the UHF/SHF region, never look into the open end of an activated length of waveguide or microwave feed-horn antenna or point it toward anyone. (If you do, you may be exposing your eyes to more than the maximum permissible exposure level of RF radiation.) In general, **make sure your antenna will be in a location where no one can get near it** when you are transmitting.

T0C05 How should you position the antenna of a hand-held transceiver while you are transmitting?

A. Away from your head and away from others
B. Towards the station you are contacting
C. Away from the station you are contacting
D. Down to bounce the signal off the ground

A   With hand-held transceivers, **keep the antenna away from your head and away from others**. Use the lowest power possible to maintain communications. If possible, use a separate microphone and hold the rig as far away from you as possible. This will reduce your exposure to the RF energy.

T0C06 Why should your antennas be located so that no one can touch them while you are transmitting?

A. Touching the antenna might cause television interference
B. Touching the antenna might cause RF burns
C. Touching the antenna might cause it to radiate harmonics
D. Touching the antenna might cause it to go into self-oscillation

B   Since the early days of radio we have known that RF energy can cause injuries by heating body tissue. Anyone who has ever touched an energized antenna or an open-wire feed line and received an **RF burn** will agree that this type of injury can be quite painful.

T0C07 For the lowest RF radiation exposure to passengers, where would you mount your mobile antenna?

A. On the trunk lid
B. On the roof
C. On a front fender opposite the broadcast radio antenna
D. On one side of the rear bumper

B   Mobile operations require some special considerations. For example, you should try to **mount the antenna in the center of the metal roof of your vehicle**, if possible. This will use the metal body of the vehicle as an RF shield to protect people inside the car. Glass-mounted antennas can result in higher exposure levels, as can antennas mounted on a trunk lid or front fender. Glass does not form an RF shield.

TOC08 What should you do for safety before removing the shielding on a UHF power amplifier?

    A. Make sure all RF screens are in place at the antenna feed line
    B. Make sure the antenna feed line is properly grounded
    C. Make sure the amplifier cannot accidentally be turned on
    D. Make sure that RF leakage filters are connected

    C   Don't operate RF power amplifiers or transmitters with the covers or shielding removed. This practice helps you avoid both electric shock hazards and RF safety hazards. A safety interlock prevents the gear from being turned on accidentally while the shielding is off. (If your equipment does not have such a safety interlock you should take other steps to **ensure that the amplifier cannot be turned on accidentally**.) This is especially important for VHF and UHF equipment. When reassembling transmitting equipment, replace all the screws that hold the RF compartment shielding in place. Tighten all the screws securely before applying power to the equipment.

TOC09 Why might mobile transceivers produce less RF radiation exposure than hand-held transceivers in mobile operations?

    A. They do not produce less exposure because they usually have higher power levels.
    B. They have a higher duty cycle
    C. When mounted on a metal vehicle roof, mobile antennas are generally well shielded from vehicle occupants
    D. Larger transmitters dissipate heat and energy more readily

    C   Although mobile transceivers usually transmit with higher power levels than hand-held radios, the mobile unit often produces less RF radiation exposure. This is because **an antenna mounted on a metal vehicle roof is generally well shielded from vehicle occupants**. The duty cycle of such operation is also generally low.

TOC10 What are some reasons you should never operate a power amplifier unless its covers are in place?

A. To maintain the required high operating temperatures of the equipment and reduce RF radiation exposure

B. To reduce the risk of shock from high voltages and reduce RF radiation exposure

C. To ensure that the amplifier will go into self oscillation and to minimize the effects of stray capacitance

D. To minimize the effects of stray inductance and to reduce the risk of shock from high voltages

B   Don't operate RF power amplifiers or transmitters with the covers or shielding removed. This practice **helps you avoid both electric shock hazards and RF safety hazards**. A safety interlock prevents the gear from being turned on accidentally while the shielding is off. (If your equipment does not have such a safety interlock you should take other steps to ensure that the amplifier cannot be turned on accidentally.) This is especially important for VHF and UHF equipment. When reassembling transmitting equipment, replace all the screws that hold the RF compartment shielding in place. Tighten all the screws securely before applying power to the equipment.

About the ARRL

The seed for Amateur Radio was planted in the 1890s, when Guglielmo Marconi began his experiments in wireless telegraphy. Soon he was joined by dozens, then hundreds, of others who were enthusiastic about sending and receiving messages through the air—some with a commercial interest, but others solely out of a love for this new communications medium. The United States government began licensing Amateur Radio operators in 1912.

By 1914, there were thousands of Amateur Radio operators—hams—in the United States. Hiram Percy Maxim, a leading Hartford, Connecticut, inventor and industrialist saw the need for an organization to band together this fledgling group of radio experimenters. In May 1914 he founded the American Radio Relay League (ARRL) to meet that need.

Today ARRL, with approximately 170,000 members, is the largest organization of radio amateurs in the United States. The ARRL is a not-for-profit organization that:

- promotes interest in Amateur Radio communications and experimentation
- represents US radio amateurs in legislative matters, and
- maintains fraternalism and a high standard of conduct among Amateur Radio operators.

At ARRL headquarters in the Hartford suburb of Newington, the staff helps serve the needs of members. ARRL is also International Secretariat for the International Amateur Radio Union, which is made up of similar societies in 150 countries around the world.

ARRL publishes the monthly journal QST, as well as newsletters and many publications covering all aspects of Amateur Radio. Its headquarters station, W1AW, transmits Morse code practice sessions and bulletins of interest to radio amateurs. The ARRL also coordinates an extensive field organization, which includes volunteers who provide technical information and other support for radio amateurs as well as communications for public-service activities. ARRL also represents US amateurs with the Federal Communications Commission and other government agencies in the US and abroad.

Membership in ARRL means much more than receiving *QST* each month. In addition to the services already described, ARRL offers membership services on a personal level, such as the ARRL Volunteer Examiner Coordinator Program and a QSL bureau.

Full ARRL membership (available only to licensed radio amateurs in the US) gives you a voice in how the affairs of the organization are governed. ARRL policy is set by a Board of Directors (one from each of 15 Divisions). Each year, one-third of the ARRL Board of Directors stands for election by the full members they represent. The day-to-day operation of ARRL HQ is managed by an Executive Vice President and a Chief Financial Officer.

No matter what aspect of Amateur Radio attracts you, ARRL membership is relevant and important. There would be no Amateur Radio as we know it today were it not for the ARRL. We would be happy to welcome you as a member! (An Amateur Radio license is not required for Associate membership.) For more information about ARRL and answers to any questions you may have about Amateur Radio, write or call:

ARRL—The national association for Amateur Radio
225 Main Street
Newington, CT 06111-1494
(860) 594-0200

Prospective new amateurs call:
800-32-NEW HAM (800-326-3942)
You can also contact us via e-mail at **newham@arrl.org**
Or check out *ARRLWeb* at **http://www.arrl.org/**

Notes

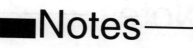

Notes

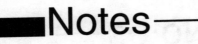

Notes

Notes

Notes

Notes

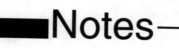

Notes

FEEDBACK

Please use this form to give us your comments on this book and what you'd like to see in future editions, or e-mail us at **pubsfdbk@arrl.org** (publications feedback). If you use e-mail, please include your name, call, e-mail address and the book title, edition and printing in the body of your message. Also indicate whether or not you are an ARRL member.

Please check the box that best answers these questions:
How well did this book prepare you for your exam?
 ☐ Very Well   ☐ Fairly Well   ☐ Not Very Well
Which exam did you take (or will you be taking)?
 ☐ Technician   ☐ Technician with code   ☐ General
Did you pass?   ☐ Yes   ☐ No
Do you expect to learn Morse code some time? ☐ Yes   ☐ No   ☐ Already know code
Where did you purchase this book?
 ☐ From ARRL directly   ☐ From an ARRL dealer

Is there a dealer who carries ARRL publications within:
 ☐ 5 miles   ☐ 15 miles   ☐ 30 miles of your location? ☐ Not sure.

licensed, what is your license class? _____

_____

Name _____   ARRL member? ☐ Yes ☐ No
_____ Call Sign_____
Address _____
 State/Province, ZIP/Postal Code _____
Daytime Phone (   ) _____ Age _____ E-mail _____
licensed, how long?_____
Other hobbies_____
_____
Occupation _____

For ARRL use only	T Q&A
Edition	2 3 4 5 6 7 8 9 10 11
Printing	1 2 3 4 5 6 7 8 9 10 11

From _____

_____

_____

EDITOR, THE ARRL'S TECH Q&A
ARRL—THE NATIONAL ASSOCIATION FOR AMATEUR RADIO
225 MAIN STREET
NEWINGTON CT 06111-1494

· please fold and tape ·